My First Numbers

Introduction to Math — Level K

1st Printing, 2024.

Printed in the United States of America.

ISBN: 978-1-954745-71-1

Cover Design: Justin Turley
Interior Design: Sarah Lee Bryant

Published by:
Generations
19039 Plaza Drive Ste 210
Parker, Colorado 80134
Generations.org

For more information on this and other titles from Generations, visit Generations.org or call (888) 389-9080.

My First Numbers

Introduction to Math – Level K

Creators and Writers
Amy Cool (BS, Business Administration; MS, Secondary Education)
Kevin Swanson (BS, Mechanical Engineering; MDiv)

Editors
Tammy Sechrist (BS, Biology)
Gina Mast (BS, Biology & Psychology)

Generations
PASSING ON THE FAITH

Contents

Your Math Toolbox

Get ready for your journey in math by gathering the following helpful tools.

Pencil

Paper

Crayons

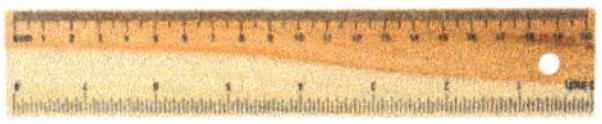

Ruler

Dice

Play Dough

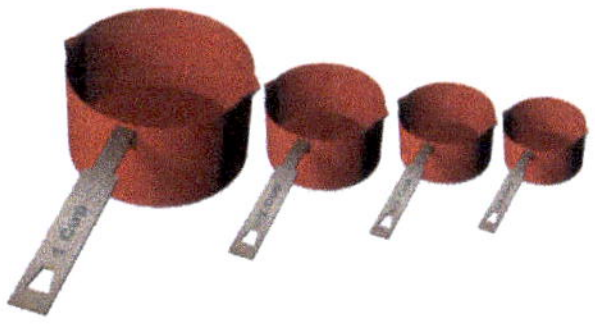

Measuring Cups

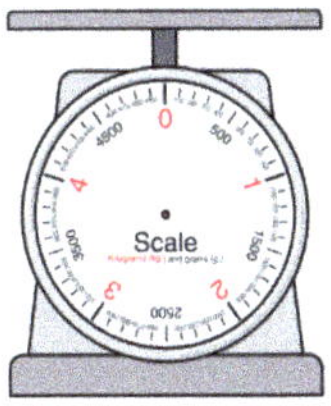

Kitchen Scale

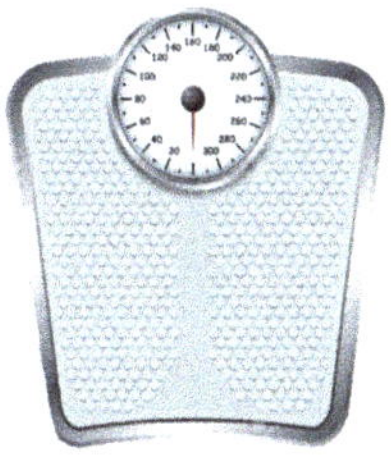

Bathroom Scale

To the Parent/Teacher

Your Child's First Math Book

This is meant to be a child's first introduction to math, created by a team of homeschool grads, experienced homeschool moms, and math-philes. But mainly, we are "God-philes"—folks who love God's Word, God's works, and God's thoughts to think after Him. This is our motivation as we work to produce a new approach to math for the Christian family. We hope you and your little learners will be blessed by the unique features in this basic math course.

Unique Features

God-Centered Focus. First and foremost, our curriculum is committed to retaining a God-centered focus. We remind the student that math comes from God. Math is God's big idea. Math helps us to understand God's world which He created. This created reality is His idea, and math helps us to measure, to study, to quantify, and to qualify God's world.

We also aim to keep the Scriptures prominent throughout the math course. Our goal is to help you implement the Deuteronomy 6:7-9 mandate, diligently teaching your children God's Word when you sit in your house, walk by the way, when you lie down, and when you rise up. Scripture is to be a "frontlet" before their eyes, integrated into every subject they study, including math.

God-Glorifying Content. Proverbs 1:7 gives us the beginning of knowledge as the fear of God. The objective behind all learning should be to know God, and to learn more of His awesome power, wisdom, and goodness. Learning must never increase the student's pride in his own knowledge, or in his ability to obtain knowledge. But rather, all learning should humble us all before the infinite wisdom of God—and bring us to wonder and praise.

Christian Worldview Built-in. Elements of a Christian worldview have been incorporated into the lessons—some of the principles may be more obvious than others. Family life, church life, honor for parents, honesty, and love for brothers and sisters are principles that are inculcated throughout.

Prayer and Faith Lessons. The time that you spend with your child will be habit forming: math habits, learning habits, and life habits. Beginning with prayer is a wonderful habit to impart: a simple confession of our dependence on God, a small acknowledgement of God's greatness, a sincere giving of thanks for the opportunity to learn, and a commitment to seek His kingdom. . . even in the context of math.

The Concrete-Abstract-Concrete Method. Math concepts are best introduced first in the concrete: by the use of manipulatives, by observing natural phenomena in creation, and by the use of real photos of God's world. This is followed up with the abstract math equations. The final step in the learning process takes math into the kitchen, the workplace, or the game room, applying math in day-to-day life. Thus, the optimum math learning experience will incorporate the concrete-abstract-concrete method.

Life Integration. Throughout the Generations math courses, students are encouraged to use math as a tool to steward God's creation, to take proper dominion, and to serve others. Knowledge is meant to be integrated into life, such that the learner becomes a "doer of the Word and not a hearer only."

Math Demonstrated in Creation. Math shows God's wisdom, organization, and unchangeable nature—His patterns. We see God's attributes in all His creation. This course uses beautiful photos to point to God's design in creation whenever possible.

How to Use this Course

1. This course includes 91 lessons, which may be spread over a 30 to 36-week school year, averaging about 3 lessons per week. The workload is appropriately laid out for the kindergarten level—adjusted to about 60% of the load that would ordinarily be assigned to a first-grade student. A K-level student following a 5-day-a-week plan would finish these lessons in 18 weeks. The parent/teacher is free to schedule these lessons as needed, depending on time available and the capabilities of the child.
2. Each chapter typically includes a mix of lessons with new material and review of prior material. New lessons are varied and presented in an engaging, wonder-filled, enthusiastic way to immerse the child in the wonderful world of mathematics.
3. For the parent/teacher's convenience, everything needed is contained in one book—relieving the busy mom from having to juggle multiple books and consuming extra time preparing lessons. An answer key is also included in the back of the book. The simple, creative, and (ideally) engaging explanations are designed for a 5-year-old comprehension level and are intended to be read aloud.

4. Begin each day with prayer. We have suggested a prayer for each lesson, but the teacher/parent/student may provide their own.
5. This course is meant to be accompanied by the use of flash cards. God has given young children an amazing ability to memorize! Scales are the foundation for learning music. Drills are the fundamentals for athletes. Basic addition and subtraction skills are essential for engaging in the art and science of mathematics.
6. First introductions to math concepts are best presented with manipulatives. The colored sheet of blocks at the end of the text may be glued to card stock and cut out for this purpose.

We Need Your Feedback

Feel free to offer any comments and suggestions to our publishing team by emailing *mail@generations.org*. We appreciate your input!

> And these words which I command you today shall be in your heart. You shall teach them diligently to your children, and shall talk of them when you sit in your house, when you walk by the way, when you lie down, and when you rise up. You shall bind them as a sign on your hand, and they shall be as frontlets between your eyes. You shall write them on the doorposts of your house and on your gates. (Deuteronomy 6:7-9)

Throughout this book, there are suggested videos, games, and crafts that you can access at Generations.org/kmath, or use this QR code!

Zero is nothing at all.

One sun.

Two people.

Three birds.

In the following picture. . .

Which pictures have only 1 of something? Write a "1" in the blank under those pictures.

Which pictures have 2 of something? Write a "2" in the blank under those pictures.

Which pictures have 3 of something? Write a "3" in the blank under those pictures.

Let's Play Dizzy-Bat Race!

If you have a baseball bat, take it outside and hold it so one end is on the ground. Lean over and put your head down on the top end of the bat. Run in a circle around the bat so that you are spinning. Count each time you go around. Can you go around 3 times? After spinning, run across your yard. What happens as you run?

LESSON 4

Let's Learn about God's Numbers from 4 to 7!

God, thank You for giving us the Bible so that we can learn more about You and about the world You made. Thank You for making my mind so I can learn about your world! Help me to learn more today from your world and your Word! Amen.

Memory Work — *Practice counting from 1 to 7, forward and backward.*

Trace each number three times. Then write the number yourself.

0	0	0	0	
1	1	1	1	
2	2	2	2	
3	3	3	3	

God made insects too!

Count the ants and ladybugs.

Can you find 6 differences between these pictures?

Color Fun!

Color 4 beautiful butterflies blue! Color 3 beautiful butterflies red! How many butterflies did you color?

How many people are in your family? How many pets do you have?

Draw a picture of your family.

Draw a picture of your pets or other animals you like.

LESSON 5 Practice with God's Numbers from 4 to 7!

Our Loving Father, thank You for making a way for us to know You. Your Son Jesus came so we can have a home for eternity! Amen.

Memory Work — *Practice counting from 1 to 7 backward and forward.*

> Then God blessed the seventh day and sanctified it, because in it He rested from all His work which God had created and made. (Genesis 2:3)

Our amazing God created everything in only a week! A week is 7 days! Let's count the numbers 1 to 7. We can learn the things God made each day of that week:

Day 1 – light

Day 2 – sky

Day 3 – dry land, seas, plants, and trees

Day 4 – sun, moon, and stars

Day 5 – creatures in the sea and creatures that fly

Day 6 – animals and man

Day 7 – a day of rest!

Count forward and backward!

1, 2, 3, 4, 5, 6, 7

7, 6, 5, 4, 3, 2, 1

Trace all the numbers from 1 to 7 you can find in this picture.

Count these things. Write how many you find below each picture.

Let's Explore Nature!

Go outside and try to collect groups of 4, 5, 6, and 7 of the same kinds of things. Some examples could be rocks, shells, pinecones, dandelions, or snowballs. These will be different depending on where you live!

Sign Language Fun!

Some people who cannot hear or speak learn to use their hands to speak. They learn **sign language**. Use the key to find the letters of the three hand shapes below. They are spelling the word that tells who created you.

Key

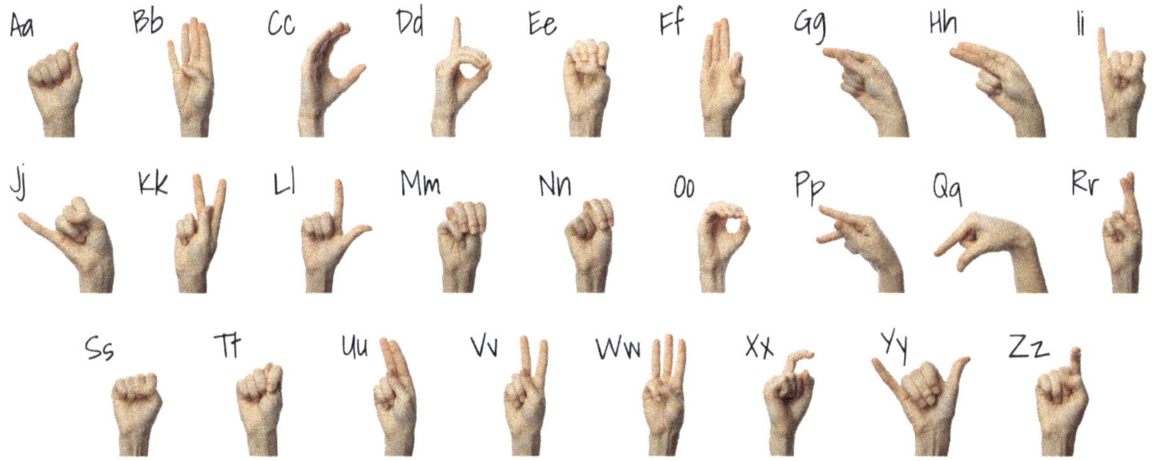

How many dots can you count on each of these dice? Write the number in each box.

How many dots can you count altogether on these two dice?

Family Game Time!

Play the game Heads Up, Seven Up with your family. This game can be played with any number of people. If you don't know the rules you can find them at game #1 at **Generations.org/kmath**.

LESSON 7

Practice with God's Numbers from 0 to 7!

Lord, You provide us with food to make us grow. Help us to be thankful for food and math!

Memory Work — *Practice counting from 1 to 7, forward and backward.*

How many apples are in the bowls?

How many cats are there?

Color the vegetables from 0 to 7. Which ones are your favorite?

Let's Watch!

You haven't learned to write your numbers all the way up to 10 yet. But you can practice saying these numbers with your parent/teacher as you watch video #2 here: **Generations.org/kmath**.

LESSON 8 Let's Learn about God's Numbers from 0 to 8!

God, thank You for numbers. Help us to use them for Your glory. Amen.

Memory Work — *Practice counting from 1 to 8, forward and backward.*

Let's read this Bible verse first, before we count to 8!

> ... In the days of Noah ... 8 persons were brought safely through water. (1 Peter 3:20 ESV)

God saved 8 people from the flood waters a long time ago. He saved Noah. He saved Noah's wife, and the rest of Noah's family. There were 8 people on the ark. Can you say, "8"? "8 people!"

Can you draw 8 people on the ark?

Practice writing these numbers: 5, 6, 7, and 8.

5	5	5	5	
6	6	6	6	
7	7	7	7	
8	8	8	8	

Draw lines to match the number with the picture that shows that many flowers.

1

2

3

4

5

6

7

8

Let's color by number!

1

2

3

4

5

6

7

8

LESSON 9 Let's Learn about God's Numbers from 8 to 10!

Creator God, You made everything with a purpose. Help us to put our 10 fingers together in prayer to praise You. Amen.

Memory Work — *Practice counting from 1 to 10.*

Today we are going to learn all ten numbers. But let's read this Bible story first.

> Now it happened as [Jesus] went to Jerusalem that He passed through the midst of Samaria and Galilee. Then as He entered a certain village, there met Him ten men who were lepers, who stood afar off. And they lifted up their voices and said, "Jesus, Master, have mercy on us!"
>
> So when He saw them, He said to them, "Go, show yourselves to the priests." And so it was that as they went, they were cleansed.
>
> And one of them, when he saw that he was healed, returned, and with a loud voice glorified God, and fell down on his face at His feet, giving Him thanks. And he was a Samaritan.
>
> So Jesus answered and said, "Were there not ten cleansed? But where are the nine? Were there not any found who returned to give glory to God except this foreigner?" And He said to him, "Arise, go your way. Your faith has made you well." (Luke 17:11-19)

Now that we have read the Bible, let's talk about what happened in this story. Jesus healed ten men who had a terrible sickness. How many is ten? Hold up your ten fingers. That's right! Jesus healed ten lepers. Later, only one came back to say, "Thank You" to Jesus. Hold up one finger. Now, where were the other 9? They never came back to say, "Thank You" to Jesus.

Let's review our numbers once more, but this time we will add two more numbers.

1, 2, 3, 4, 5, 6, 7, 8, 9, 10

Ten is a complete number. We each have 10 fingers and 10 toes! Every finger and toe was made by God! Let's practice counting to 10 one more time using our fingers. Look at this number carefully:

10

You can see there are two numbers side by side. Ten is written with a 1 and a 0. We call this ten.

Time to Create and Sing!

Let's sing the song "Five Little Monkeys Jumping on the Bed." But first, make some monkeys by coloring pictures or by making them out of craft dough. Remember, the monkeys in the song got hurt. This is why we don't jump on our beds! If you don't know the song, you can watch video #3 at **Generations.org/kmath**.

Practice writing these numbers too.

0 1

2 3

4 5

6

7

8

9

10

Draw a group of 8 fish.

Draw a group of 9 pieces of candy.

Draw a group of 10 apples.

Game Time! Let's play hopscotch!

A parent or teacher can help you draw out a hopscotch course with chalk on your driveway. If you don't have chalk, you can use numbered sheets of paper or cardboard, one laid down for each square. Practice counting as you hop. Try to hop different ways—on one foot, on the other foot, even backward!

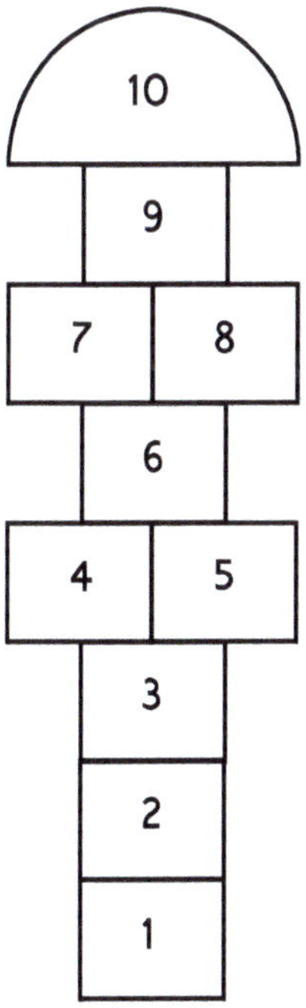

LESSON 10

Practice with God's Numbers From 1 to 10!

God, help us have fun with math! Help us use math to learn more about what You've created.

Memory Work — *Practice counting from 1 to 10, backward and forward.*

Let's practice our numbers 1 to 10 today!

Circle the right number for each of the following exercises.

4 8 9

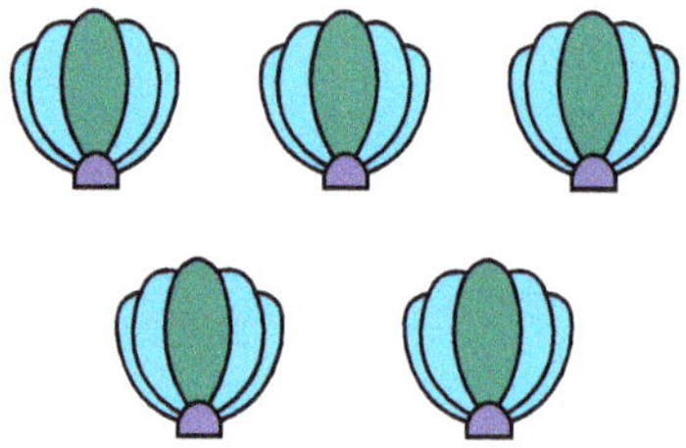

5 7 6

2 3 4

1 3 5

10 8 7

9 10 7

Count the objects, and match with the correct number by drawing lines.

6

7 2

1 8 5

4 3

10

9

START
FINISH

Activity

Parent/Teacher: Make a spinner and play a game using it. You may want to use the game board to the left. You can cut out this wheel and glue it to cardboard, then use a brass fastener or pencil point for a pivot and a paperclip for a spinner. Refer to game #2 for a creative way to employ spinners in teaching: **Generations.org/kmath**.

CHAPTER 2

God's Colorful Creation

LESSON 11

Primary Colors

Our loving Father, thank You for the beauty around us. You made a colorful world for us to see. Help us to remember that You are the One who painted the sky. Amen.

Memory Work — *Practice recognizing primary colors in the home.*

> He answered and said to them, "When it is evening you say, 'It will be fair weather, for the sky is red.'" (Matthew 16:2)

Isn't God amazing? He created so many beautiful things for us to look at. We see colors all around us.

Primary colors are the only colors that can't be made by mixing other colors. Instead, all other colors are made by mixing certain primary colors! The primary colors are **red**, **blue**, and **yellow**.

God made many of the beautiful foods we eat primary colors.

Color the bananas yellow!

Draw It!

Go outside and draw a picture of something beautiful that God created. Do you see any primary colors?

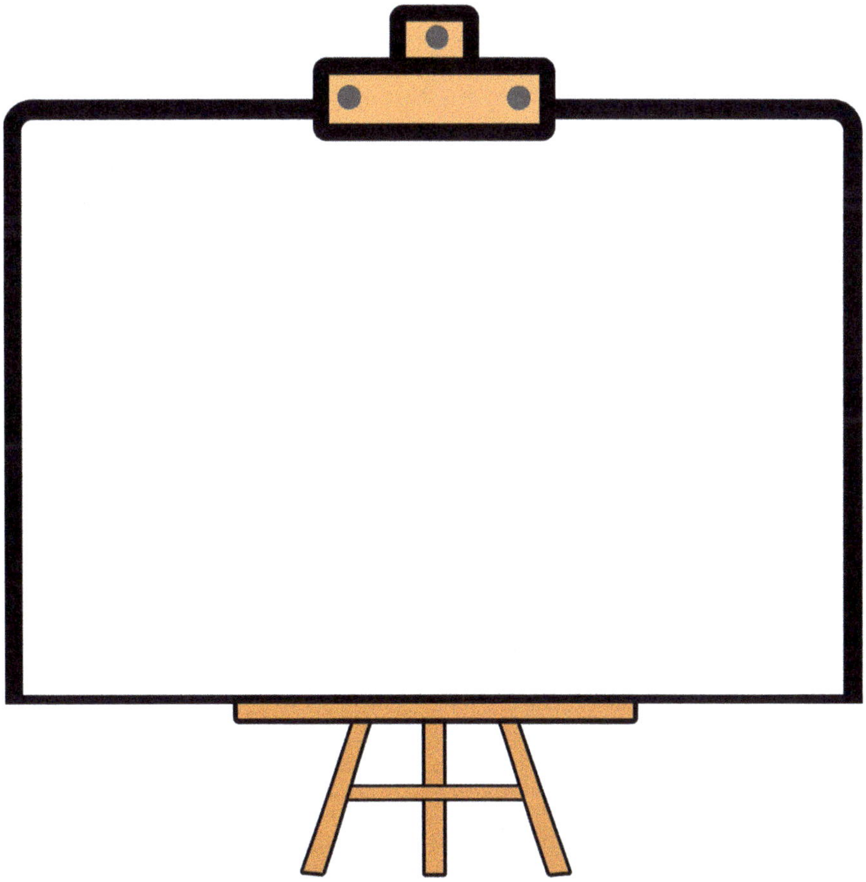

Color the apple red and the book blue!

Trace the words using the right color.

Red

Yellow

Blue

Count how many!

How many **red** balls do you see?

______

How many **yellow** balls do you see?

______

How many **blue** balls do you see?

______

LESSON 12 Secondary Colors

God, thank You for green pastures that feed the animals and give us room to run and play. Amen.

Memory Work — *Practice recognizing secondary colors in the home.*

God made **secondary colors**. Secondary colors can be made by mixing specific primary colors. See if you can use primary colors of paint or playdough to make these secondary colors:

Mix blue and red to make **purple**.
Mix red and yellow to make **orange**.
Mix yellow and blue to make **green**.

How many oranges do you see?

Count the round, green bushes. Do you think they grew this way, or do you think someone is taking care of them?

Color the traffic light.

Color the top circle red. Color the middle circle yellow. Color the bottom circle green. Which colors are primary and which color is secondary?

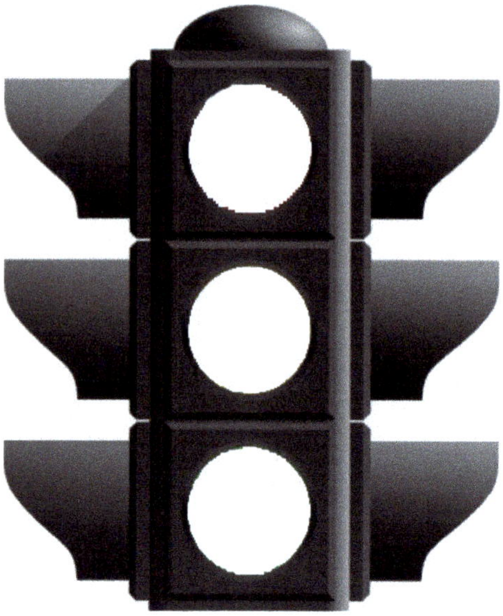

People often mark their Bibles with colors to bring them back to the verses that they love.

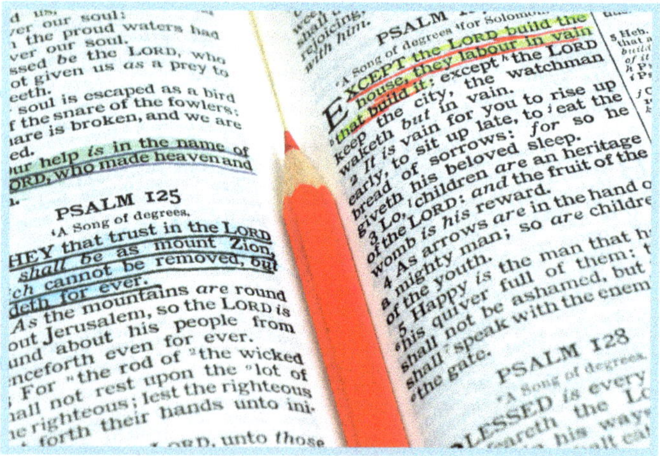

Let's Watch!

Watch video #4 to help you remember what makes primary and secondary colors: **Generations.org/kmath**.

Color the butterfly.

See if you can match the one on the bottom of the page. Can you tell which colors are primary and which are secondary?

LESSON 13

Fun with Number Lines

Father God in heaven, it is so good to know more about You every day. Thank You for the gifts You give, including these numbers! Amen.

Memory Work — *Practice counting from 1 to 10, forward and backward.*

Now that you have learned to count to 10, let's put the numbers on a line! We call this a *number line.* Can you say "number line"?

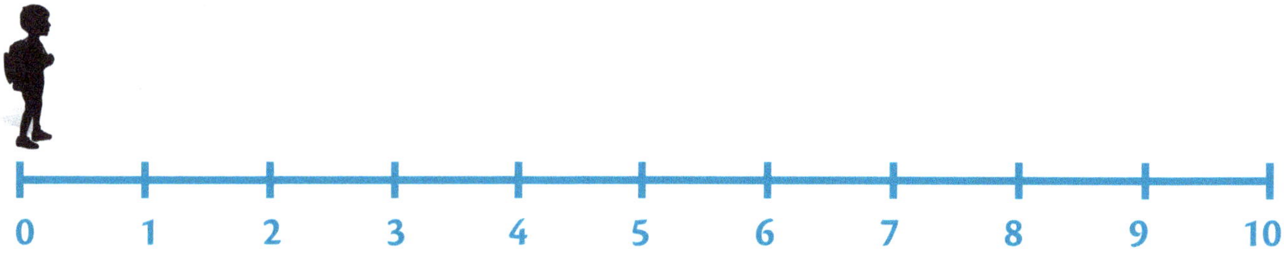

You can see that the numbers are connected. First comes 0. Then comes 1. What comes next? Let's pretend that you are starting at 0 and walking with this little boy along the line!

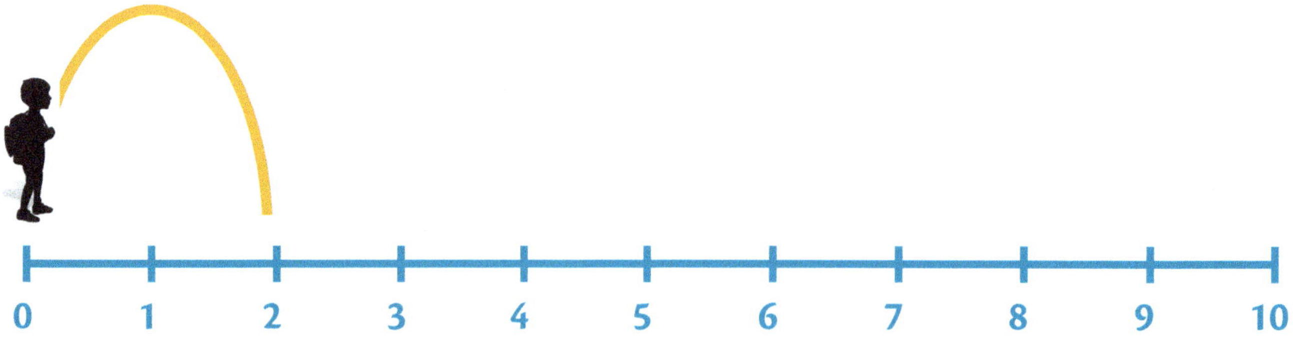

How many spaces did the boy walk? ______

How many spaces do you walk to get to 10? That's right! 10 spaces!

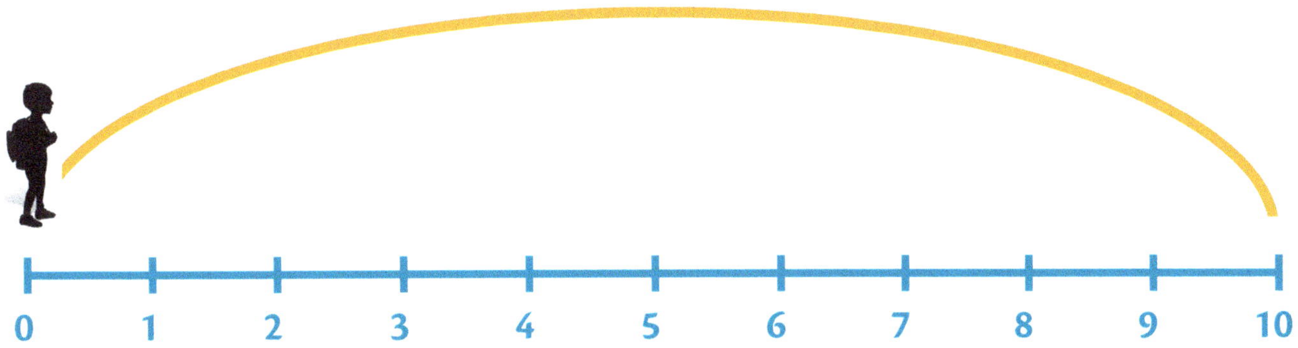

Now, instead of walking along the number line, let's pretend to jump!

Draw a jump of the right length on each number line below. In each box, write the number of number-line spaces the child jumped over.

From 7 to 8:

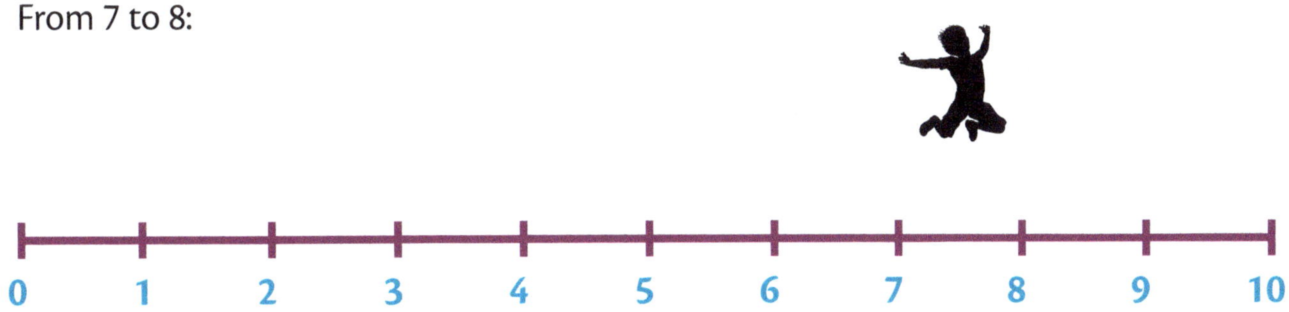

How long was this jump? How many spaces?

From 0 to 2:

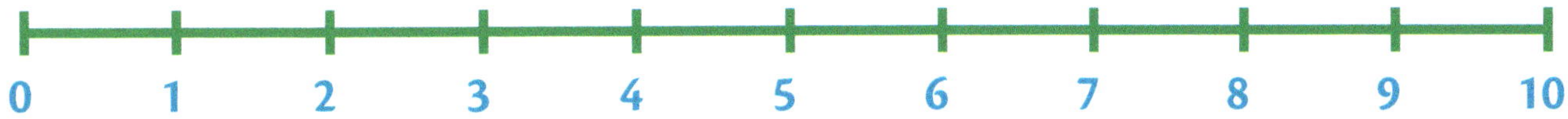

How long was this jump? How many spaces?

From 2 to 3:

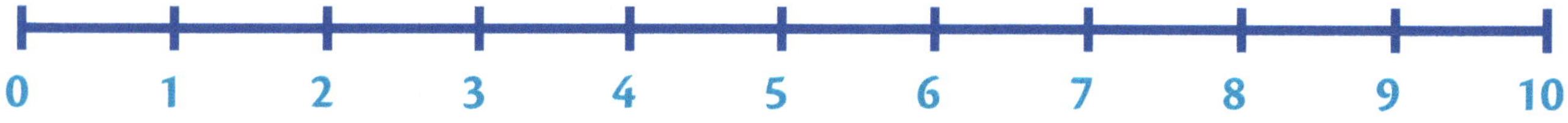

How long was this jump? How many spaces?

From 1 to 4:

How long was this jump? How many spaces?

From 3 to 10:

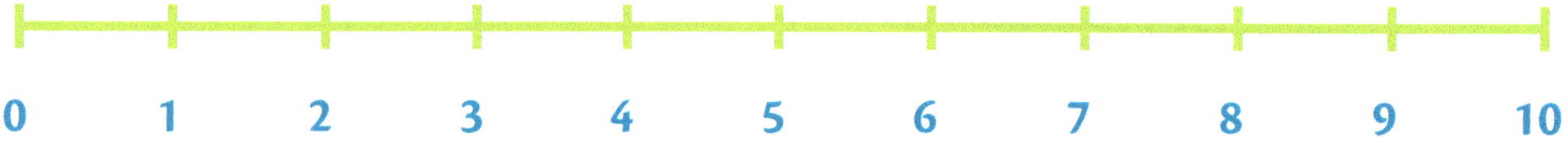

How long was this jump? How many spaces?

Let's Jump!

Gather some sidewalk chalk and find a sidewalk or driveway. Draw a number line with marks from 0 to 10. Space the marks 1 foot (30 cm) apart.

Try jumping from 0 to 2. Was that easy or hard? Can you jump from 5 to 8? How many spaces would you jump over if you jumped from 5 to 8? Take your biggest jump. How far (how many spaces) did you jump?

LESSON 14

Practice Your Numbers

God, help me to write numbers so that one day I can copy Scripture.

Memory Work — *Practice counting from 1 to 10, forward and backward.*

Trace these numbers. Count each of the things next to the chart and draw a line to match the things with their number.

1	1	1	1	1
2	2	2	2	2
3	3	3	3	3
4	4	4	4	4
5	5	5	5	5

6	6	6	6	6
7	7	7	7	7
8	8	8	8	8
9	9	9	9	9
10	10	10	10	10

LESSON 15

Adding Numbers 0, 1, and 2

Father God in heaven, thank You for adding blessings to me every day. Help me to understand math today! Amen.

Memory Work — *Practice counting from 1 to 10 forward and backward.*

First, we will read these verses from the Bible. Listen carefully.

> Now Adam knew Eve his wife, and she conceived and bore Cain, saying, "I have gotten a man with the help of the Lord." And again, she bore his brother Abel. Now Abel was a keeper of sheep, and Cain a worker of the ground. (Genesis 4:1-2)

At first there were two people—a man and his wife. Who were they? That's right! Adam and Eve. They were the first daddy and mommy. Then God gave this mommy and daddy two boys. He added the boys. Now how many people were in the family? Can you count the people in this picture? How many boys do you see? How many people are there altogether?

Now, pretend that the numbers on this number line are people. Let's start with 2 people. Put your finger on the 0 and move it to the number 2. Now we'll add 2 little boys. Move your finger 2 more spaces. What number is your finger on? That's how many people there are altogether.

We have moved from 2 to 4 on the number line. There were some people. Then, more people came.

When we put numbers of things together, it's called **adding**. Can you say "adding"?

Adding brings things together to make bigger numbers. Let's add bananas. We will start with 1 banana. You can write the number 1 under one banana. Write another number 1 under the other banana. Now we will bring the bananas together. How many bananas are there together? That's right! You can see that there are now 2!

Do you see the + picture between the and the ? When we see +, we say the word *plus*. This means we are adding the two things together. We say it this way: *1 banana plus 1 banana.*

Now do you see the = picture between the + and the ? When we see =, we say the word *equals.*

Let's say it all together! *1 banana plus 1 banana equals 2 bananas.*

Let's try adding people in a family. We'll start with a daddy and mommy. Then we'll add 2 children. Write the number of people you see under each picture.

We say it this way: *2 people plus 2 people equals 4 people*. Now you say it!

Now, try these adding exercises yourself.

Write the number of bananas on the line underneath. Can you write the + and the = too?

Let's add some cherries too!

LESSON 16 Adding More Numbers from 0 to 5!

God, we are amazed that You know everything! We praise You for sharing math with us so that we can learn about the things You have made.

Memory Work — *Practice counting 1 to 10 and recognizing primary and secondary colors.*

Parent/Teacher: For this lesson, the child may use the cutout manipulative blocks in the back of the book.

When you add numbers together, the answer is called a **sum**. Can you say "sum"?

Let's add more numbers today by adding little squares. We'll call the squares "blocks."

The smallest blocks are blue blocks made of only one square. We'll call these "blue 1-blocks." Let's add a blue 1-block and another blue 1-block together. The sum is 2 blue 1-blocks.

In your cutout blocks, you also have orange blocks. Orange blocks are the size of 2 blue blocks put together. Let's call these "orange 2-blocks."

We can see in this block picture that a blue 1-block plus another blue 1-block also equals 1 orange 2-block.

Now, let's try one more. Let's add a blue 1-block and an orange 2-block together. This makes a green 3-block. We put the blue block and the orange chunk of blocks together, and we get the green chunk of blocks. Write the number of blocks you see under each of the chunks of blocks.

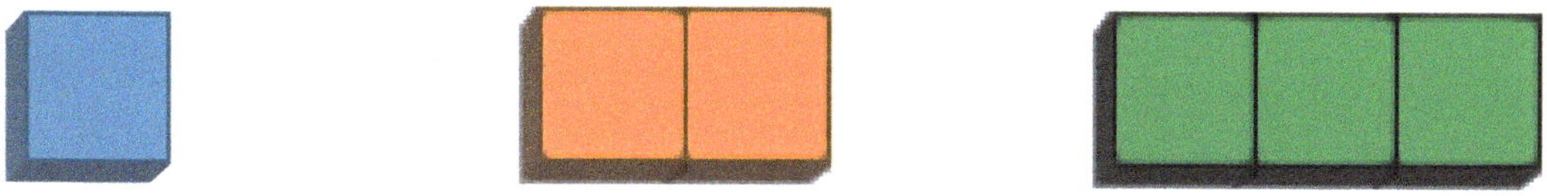

____ + ____ = ____

Let's try another one. We add a green 3-block to a blue 1-block, and we get a big pink chunk of 4 blocks! Write the number of blocks you see on each of the lines.

____ + ____ = ____

Here you can see 3 kittens with orange fur on their faces. Let's add these 3 kittens plus 2 gray kittens. How many kittens are there altogether? Can you write the number on the line?

3 + 2 = ____

Add 1 gray bunny and 2 orange bunnies. How many bunnies will there be? Write the number on the line.

____ + ____ = ____

Add these frogs together.

1 + 1 = ____

Color the pictures of these rhinos, and then add them together. How many rhinos are there altogether? Write the number on the line.

 +  = ____

Now think of the numbers. Write the sum of 2 + 2 on the line.

2 + 2 = ____

Add 3 chicks plus 0 chicks. How many are there altogether? Write the answer on the line.

 + = ____

Now think of the numbers and write the sum.

3 + 0 = ____

Let's add some fish that God made. Color 3 fish green. Color 1 fish yellow. What is the sum of 3 + 1? Write the number on the line.

3 + 1 = ____

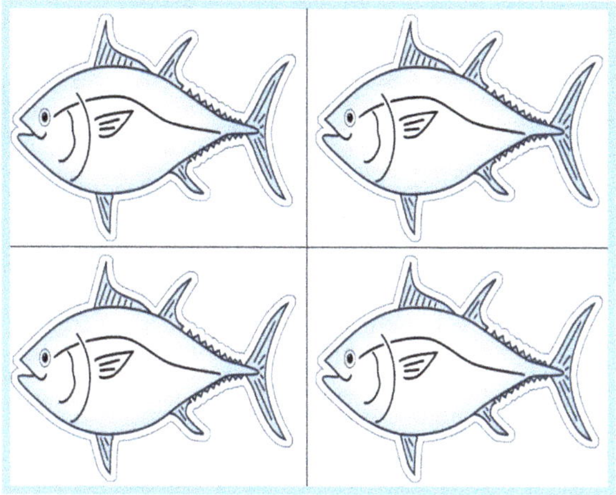

How many ears of corn will you have when you add 3 + 2?

3 + 2 = ____

Let's Exercise!

See if you can add to find the sums in this video. Do the fun exercises too! Access game #3 at **Generations.org/kmath**.

Parent/Teacher: References to "solving the equation" is the same as finding the answer, or adding to find a sum.

LESSON 17

Addition, 0 to 5!

God, we praise You for the blessings around us. Help us be thankful for all the things we have. Help us use our blessings for Your glory! Amen.

Memory Work — *Practice counting from 1 to 10 forward and backward.*

Help Dad add up his tools.

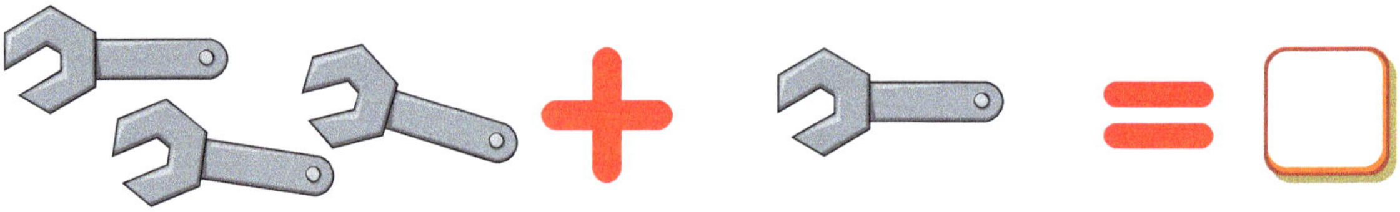

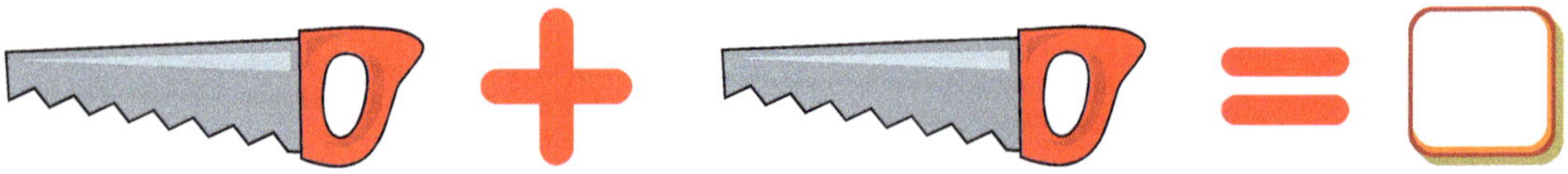

Help Mom add in the kitchen.

Add the balloons.

Explore Your Room!

Find some things in your room that you can add. Draw pictures of these things. Add them and write down the answers.

[ ] + [ ] = [ ]

Draw and add some more things from your room!

[ ] + [ ] = [ ]

LESSON 18

More Addition, 0 to 5!

God, thank You for helping us learn to add by counting. We want to take good care of the earth You made. Amen.

Memory Work — *Practice counting from 1 to 10, backward and forward.*

Count and color the animals!

○ + ○ = ○

○ + ○ = ○

○ + ○ = ○

Game Time!

Here is an online game. Add up cars... or marbles or other objects! See game #4 at **Generations.org/kmath**.

Math Maze!

Add to find the sum for each house. Then find the child whose key is that sum. Draw a line along the path between the child and his or her home. You can use your cutout blocks to help you add.

LESSON 19

Comparing Sizes and Numbers

Lord, we know that You are greater than anything we can imagine. Help us learn our math today so that we can be used by You all our life. Amen.

Memory Work — *Practice counting from 1 to 10 backward and forward and notice which numbers are bigger or smaller.*

> "This day the Lord will deliver you into my hand..." (1 Samuel 17:46a)

People like to compare things. In this picture of David and Goliath, we can see that Goliath was bigger. He was an older, trained soldier with better weapons. David was a boy. He was too young to be in Israel's army.

Goliath thought he would win. Goliath didn't realize that God was in control. God used David to defeat Goliath.

Let's Color and Compare!

Which one is bigger? Circle either the police car or the fire truck then color them.

Which one is bigger? Circle either the laptop or the desktop computer.

Which one is smaller? Circle either the basketball or the tennis ball.

Compare the Numbers!

How many are in each group? Are there more yo-yos or more wagons? Circle the bigger group.

Do you see more strawberries or more nectarines? Circle the bigger group.

LESSON 20

Compare Weights

God, You have made big things and small things. You are stronger than all of them. But You are gentle too. Help us to be happy with all You have done. Amen.

Memory Work — *Practice counting from 1 to 10 and notice witch numbers are bigger or smaller.*

So far, we have compared things by size and by number. Now, let's compare by weight. Pick up one of your shoes and one of your dad's shoes. Which one is heavier? Heavier things are harder to lift. Your dad's shoe is probably heavier than your shoe.

Which one is heavier? Circle either the toy truck or the skateboard.

Which one is lighter? Circle either the pillow or the lamp.

Put the gardening supplies below in order from lightest to heaviest. Label from 1 to 3.

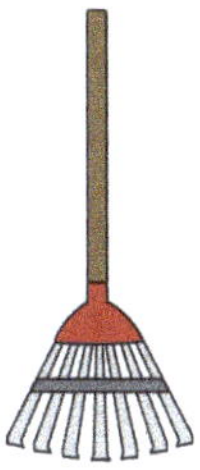

Let's Review! Practice counting and writing your numbers!

•	1	1	1	1	1
• •	2	2	2	2	2
• • •	3	3	3	3	3
• • • •	4	4	4	4	4
• • • • •	5	5	5	5	5
• • • • • •	6	6	6	6	6
• • • • • • •	7	7	7	7	7
• • • • • • • •	8	8	8	8	8
• • • • • • • • •	9	9	9	9	9
• • • • • • • • • •	10	10	10	10	10

Sing and Move!

Watch video #5 about heavy and light things. Use motions to show what is heavy or light. **Generations.org/kmath**

LESSON 21 More Comparisons

God, thank You for the gift of faith. Help our faith continue to grow in You. May others be encouraged by the faith they see in us. Amen.

Memory Work — *Practice counting from 1 to 10 and recognizing primary and secondary colors.*

Have you ever seen a mustard seed? It is very small compared to a large oak tree. Our faith may seem small like the mustard seed. We have nothing to fear if our faith is in Jesus. He is in control!

Compare the weight.

1. Which is heavier: an apple or a watermelon?
2. Which is heavier: a pine tree or a pea?
3. Which is heavier: an apple or a pine tree?
4. Which item is the heaviest? Which is the lightest?

Discovery Activity!

Get a small kitchen scale, some measuring cups, and something easy to pour, such as cereal. Guess which measurements you think will be heavier or lighter. Use the scale to see if you were right.

How many of each kind of bug do you see?

Count the number of each kind and write the number in the correct white circle.

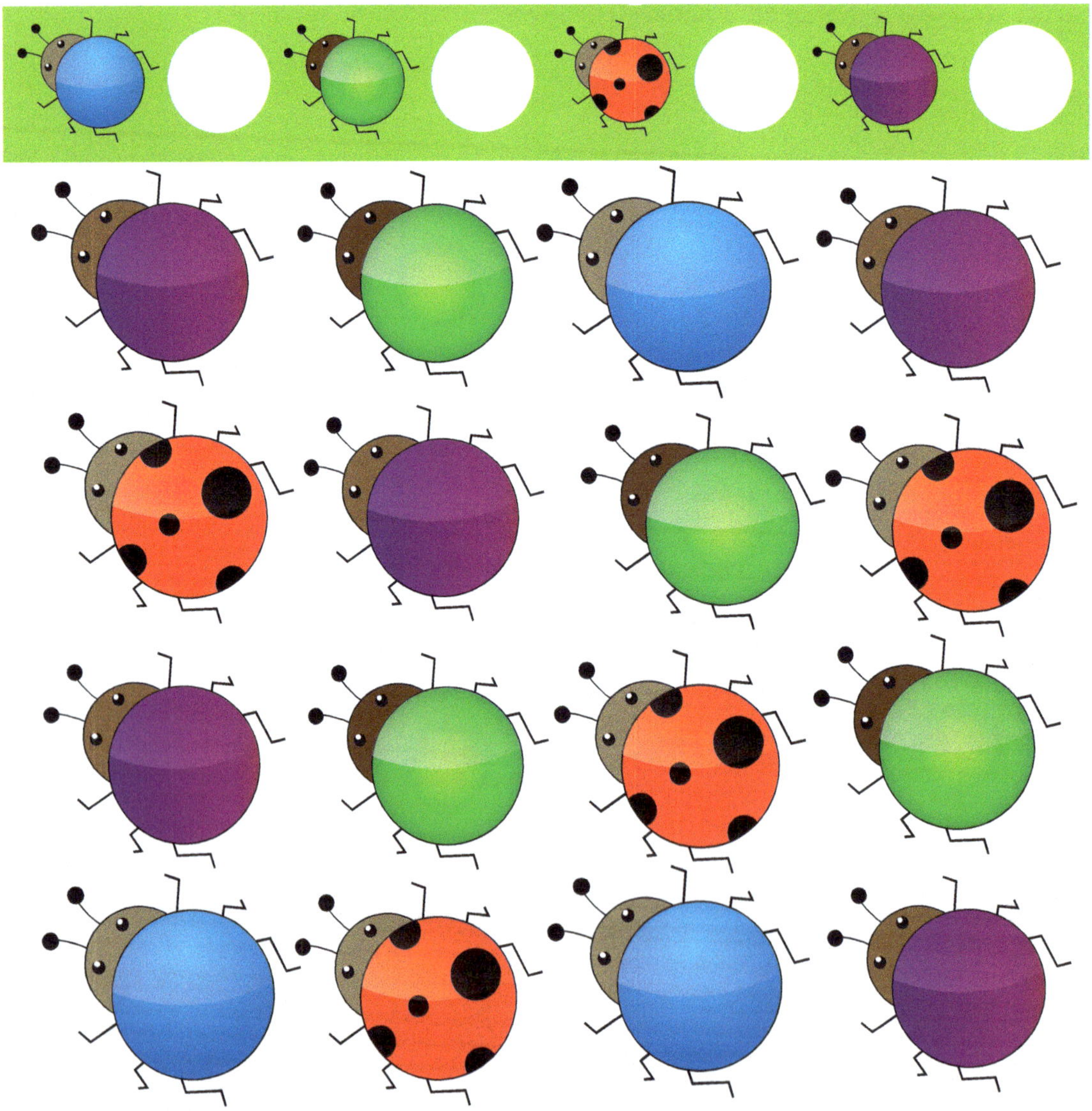

Practice counting as you connect the dots to complete the bear. Color your picture!

CHAPTER 3

God's Shapes

LESSON 22

Circles

God, Your wisdom never ends. We know You are perfect. You will help us learn math when we ask for Your help. Amen.

Memory Work — *Practice recognizing and drawing a circle.*

> Then Joshua the son of Nun called the priests and said to them, "Take up the ark of the covenant, and let seven priests bear seven trumpets of rams' horns before the ark of the Lord." And he said to the people, "Proceed, and march around the city, and let him who is armed advance before the ark of the Lord." (Joshua 6:6-7)

Shapes are everywhere in our world. A circle is a shape that is round. It has no beginning and no end. A circle can remind us of God. God had no beginning. He has always been here. God will have no end. He will always be here.

The Bible sometimes talks about shapes. The Bible talks about a circle! A man named Joshua marched in a circle around the city of Jericho. God's people marched in the circle with Joshua. After they walked around the city seven times, the walls came tumbling down! It was a miracle! Walls don't just fall on their own. God made them fall down.

God has made so many circles. Have you seen these circles before? Name them.

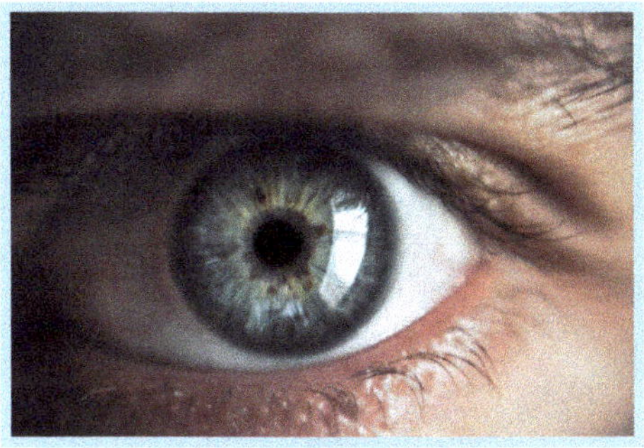

Circles are everywhere!

We can make circles too! Look at all these circles! Do you see the circles that the boat is making in the water?

The hot air balloon over you looks like a circle. Eyes have a circle inside of a circle. When you look up at a basketball net, you see a circle.

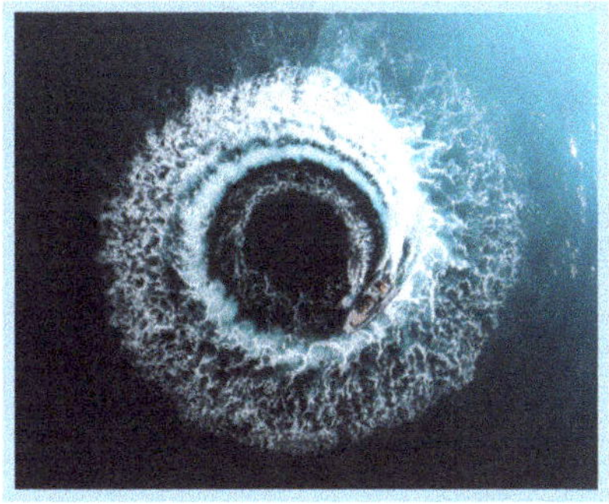

Now it's your turn to draw a circle. Draw a small circle inside the two big circles. Make sure that the dot is still in the middle of your little circle. The dot is called the **center** of the **circle**.

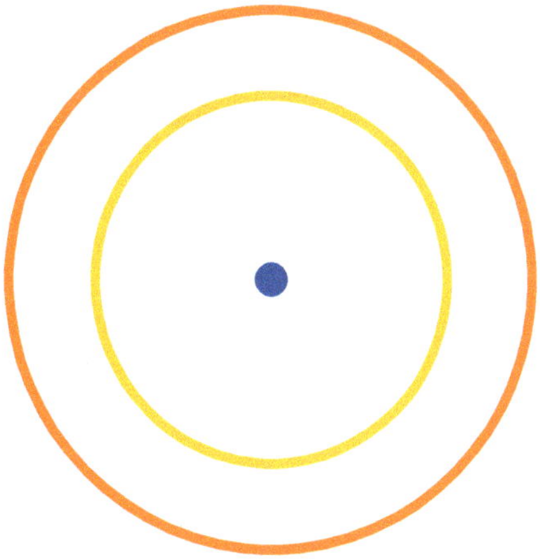

Help the girl run through the circle maze to get to the ice cream. Color the picture.

LESSON 23 More Circles

God, Your wisdom never ends. We know You are perfect. You will help us learn math when we ask for Your help. Amen.

Memory Work — *Practice recognizing and drawing a circle.*

The sun, planets, and moons that God made look like circles in the sky. Color as many of these circles as you can see in the picture.

Count the circles or wheels, and write the number in the box below each picture.

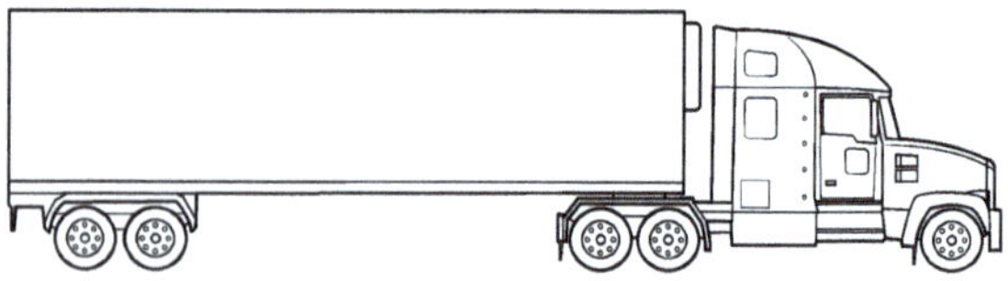

Draw circles to make gumballs inside the machine. Color the gumballs.

Let's Play!

Find some friends or family members. Take them outside to play Duck, Duck, Goose. To play, choose someone to be "It." Everyone else will be "ducks." Ask the ducks to sit in a circle. The person who is It walks around the outside of the circle. He taps each duck on the head as he passes. He counts each one: "1 duck, 2 ducks…" as he taps the ducks. Then It decides which duck he wants to call a "goose." He taps that duck on the head and yells "Goose!" Goose chases It around the circle. Quickly, It tries to get all the way around the circle and sit in Goose's spot. If It gets there without being caught, he becomes a duck and Goose has to be It. If Goose catches It, he must be It again.

LESSON 24

Squares

God, thank You for giving us sturdy shapes that we can use to build things. Amen.

Memory Work — *Practice counting from 1 to 10 and drawing a square.*

> "It [the breastplate of judgment] shall be doubled into a square: a span shall be its length, and a span shall be its width." (Exodus 28:16)

Circles are drawn with curved lines. Other shapes are drawn with straight lines. These lines might be long. These lines might be short. When we talk about how long a line is we are talking about its **length**. Can you say, "length"?

Squares are shapes that have **4 sides. The sides of squares must all be the same length.** Here is a yellow square. What is the difference between a square and a circle?

Can you see squares in these pictures below? Waffles can be square shaped. They usually have little squares inside them. Some windows, pillows, or gift boxes may also be square shaped. Where else have you seen squares?

Have you ever seen anyone sew a quilt? Quilts are often made with squares in different colors.

Can you count these squares? How many squares are in a row (across, or side to side)?

How many squares are in a column (up and down)?

Which of these squares look the same and which of them look different?

It's Game Time!

With parental or teacher supervision, you can play checkers using an online program. Access game #5 at **Generations.org/kmath**.

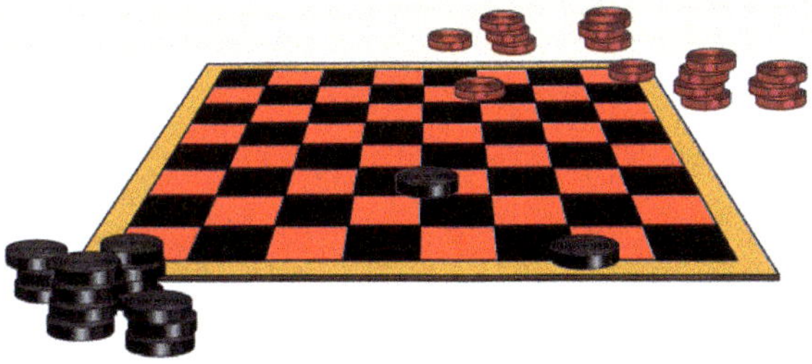

LESSON 25

More Squares

God, thank You that we can make things. Help us to enjoy seeing all the different shapes around us.

Memory Work — *Practice counting from 1 to 10 and drawing a square.*

Desserts shaped like squares may make you hungry. Brownies are often cut into square shapes after they are baked in the oven. S'mores made with graham crackers are in the shape of squares. Ask your parents if there is a sweet treat you can help make for your family.

Have you noticed squares while watching a baseball game? If the batter could draw lines between the bases as he ran, he would draw a square. Each side of that square would be 90 feet long!

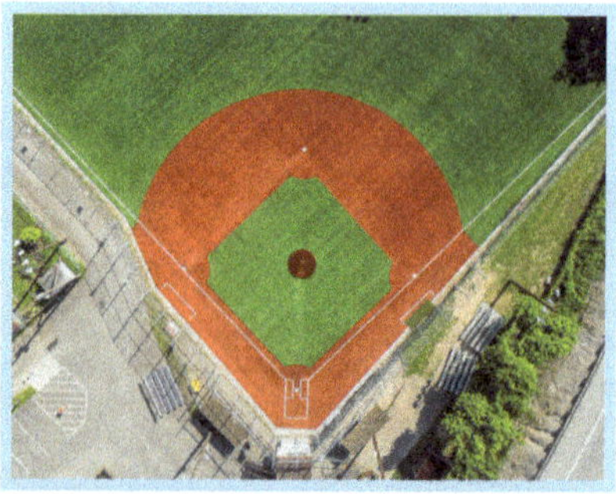

There is also a square on the backboard of a basketball hoop. Bouncing the ball off it may help you get a basket.

Create a picture using shapes you've learned about so far!

Use as many squares and circles as possible in your picture. How many did you make?

Let's Exercise!

Directions for a fun game (game #6) may be accessed at **Generations.org/kmath**.

LESSON 26

Practice Shapes and Numbers

God, help us to build things that honor You with the shapes that You have given to us.

Memory Work — *Practice drawing circles and squares.*

Count the objects. Then add them together and write the answer in the box.

___ + ___ = ___

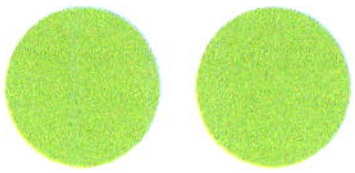

___ + ___ = ___

___ + ___ = ___

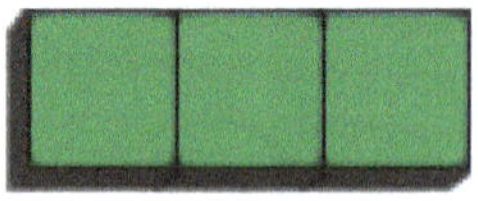

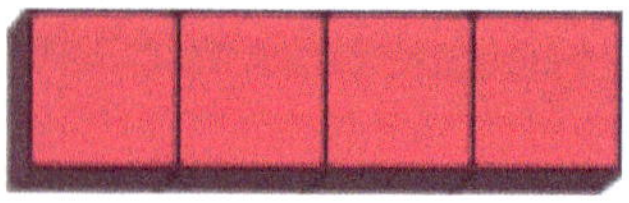

___ + ___ = ___

Let's build things with shapes, and color them too!

Color all the circles red. Color the squares blue. Color everything else green in this picture.

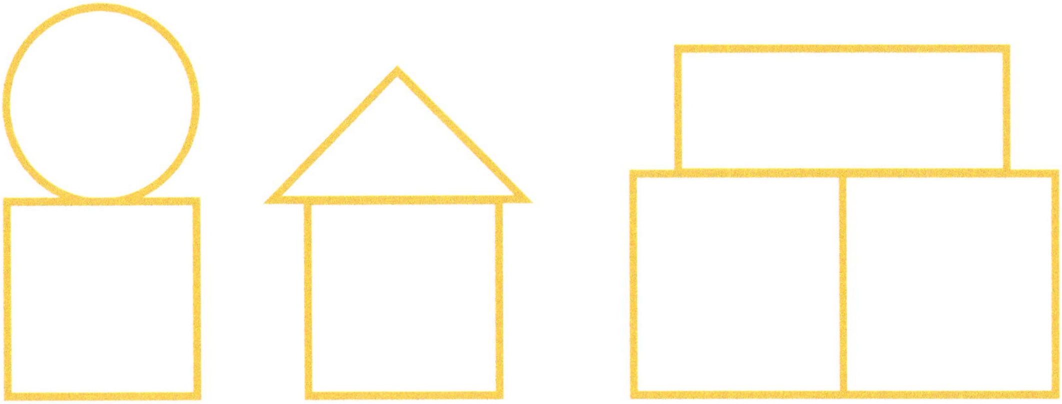

Count these! How many do you see? Write the number on the line.

Make your own drawings of shapes or objects for each of the numbers. Then, add them together and put the answer on the line.

1 + 1 = ____

2 + 2 = ____

0 + 3 = ____

LESSON 27

Rectangles

God, thank You for making a way to save Your people.

Memory Work — *Practice drawing a rectangle and a square.*

> "And they shall take some of the blood and put it on the two doorposts and on the lintel of the houses where they eat it." (Exodus 12:7)

Rectangles are shapes made with 4 straight lines. The 2 lines that are across from each other must be the same length. That means that the top and bottom lines are the same length, and the lines on the sides are the same length.

Doors are usually shaped like rectangles. God used doorposts like the one shown in the picture to save His people during Passover.

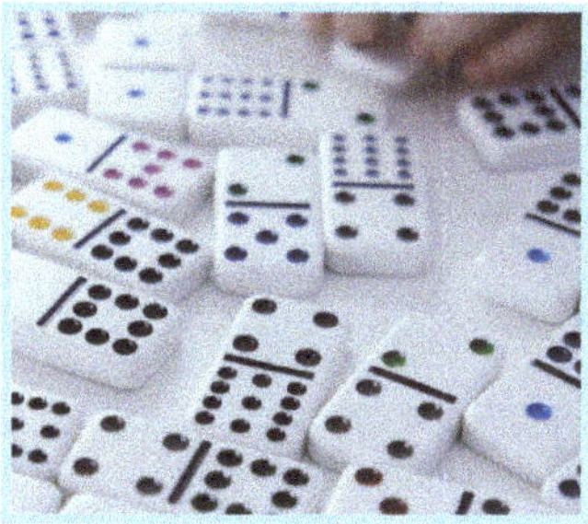

You've probably seen a lot of rectangles. The walls of houses are usually shaped like rectangles. People like to build pools in the shape of rectangles. Dominos are rectangles. If you have ever watched or played soccer, the field was probably a rectangle.

Can you tell the difference between a rectangle, a square, and a circle?

Let's count the rectangles, the circles, and the squares in this picture! Write the number you count for each shape in the boxes.

Rectangles

Circles

Squares

Here is a picture made out of rectangles. Can you count how many rectangles are drawn of each color?

Optional Activity

Make a similar picture for yourself by drawing or by using rectangles you have cut out of construction paper.

LESSON 28 More Rectangles

God, thank You for allowing us to build our relationship with You!

Memory Work — *Practice drawing a rectangle and a square.*

Let's talk about rectangles! Today, we want you to tell the difference between circles, squares, and rectangles.

What is this? Is it a square or a rectangle?

What is this? Is it a square, a circle, or a rectangle? Would you like to drink out of a cup that is shaped like a square? Why or why not?

Are these books shaped like squares or rectangles?

Look at the books in your house. Are they shaped like squares or rectangles? How would you like to read a book shaped like a circle? Or what about a book shaped like a square?

Sometimes rectangles are better than squares. And sometimes circles are better than rectangles!

Pattern Time!

Find and color the matching shapes.

Circle – purple   Rectangle – green   Square – yellow

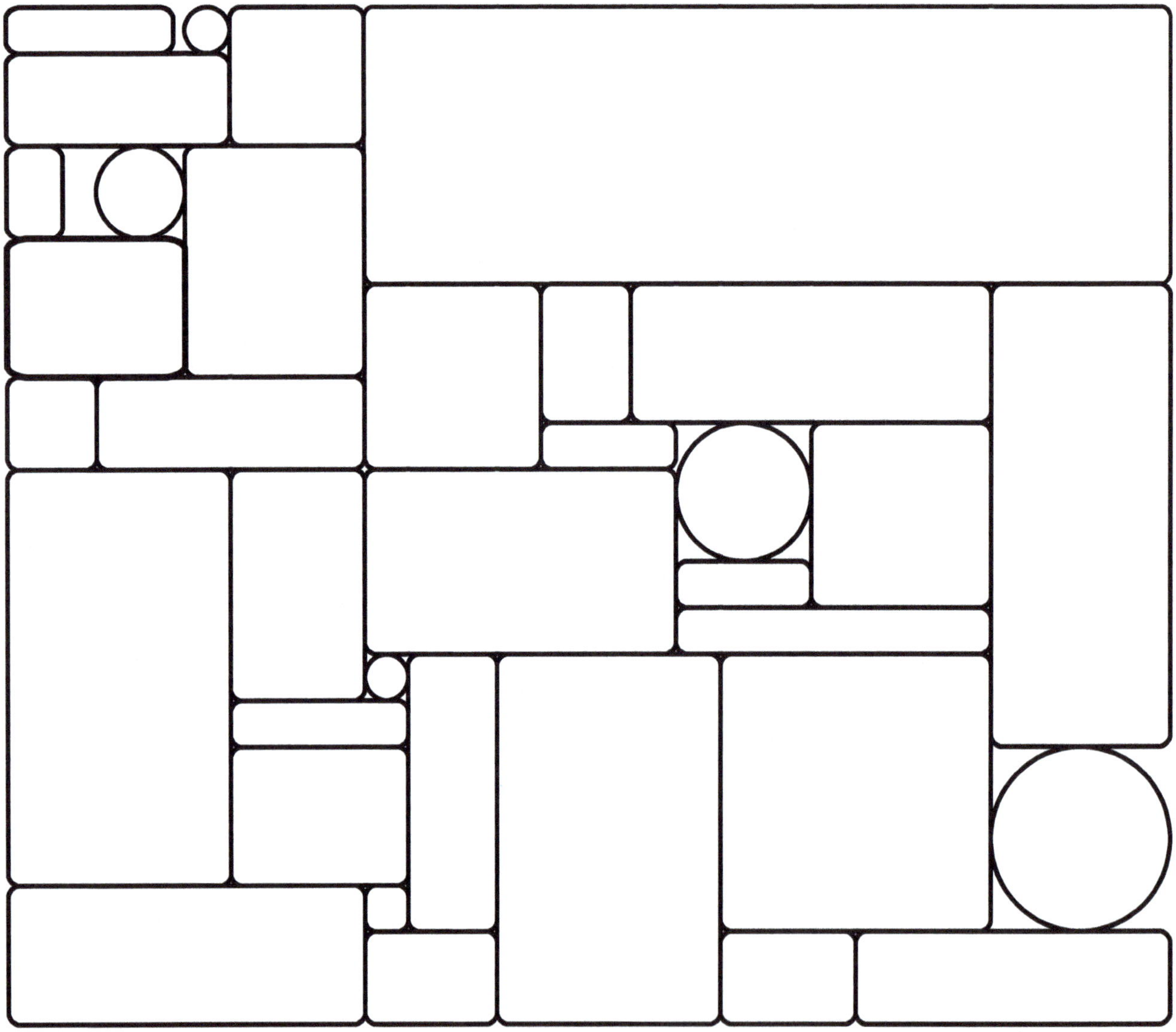

Activity

If you have LEGO® bricks, you can divide them into two piles of rectangles and squares. Try building some things with squares and other things with rectangles.

LESSON 29

Triangles

God, help us to always remember to try our best in math.

Memory Work — *Practice counting from 1 to 10 and drawing a triangle.*

Today you will learn about triangles! Squares and rectangles are made out of 4 lines connected at the corners. But triangles are made out of 3 lines connected at the corners. What is the difference between a triangle, a square, and a circle?

Did you know that there is a musical instrument called a triangle? It is in the shape of a **triangle**.

Let's Listen!

Access video #6 for a video of someone playing a triangle. Access video #7 to help you learn about the shapes of triangles: **Generations.org/kmath**.

Here are some more triangles to see. In the game of pool, the balls are often set in a triangle shape to begin playing.

This yummy piece of pizza is shaped like a triangle.

You may have seen road signs in the shape of a triangle.

Make It!

Cut out triangles on page 281 and arrange them into a boat shape like this one.

Challenge Game

Cut out the colored shapes on page 283 and see if you can arrange them to form the shadow images below.

Game #1

Game #2

LESSON 30

More Triangles

God, thank You for all the different shapes we have learned. We want to use them to see the beauty in the world You have made.

Memory Work — *Practice drawing a circle, square, rectangle, and triangle.*

Find and color all the circles, squares, triangles, and rectangles with the right colors!

Match the correct shape with the road sign.

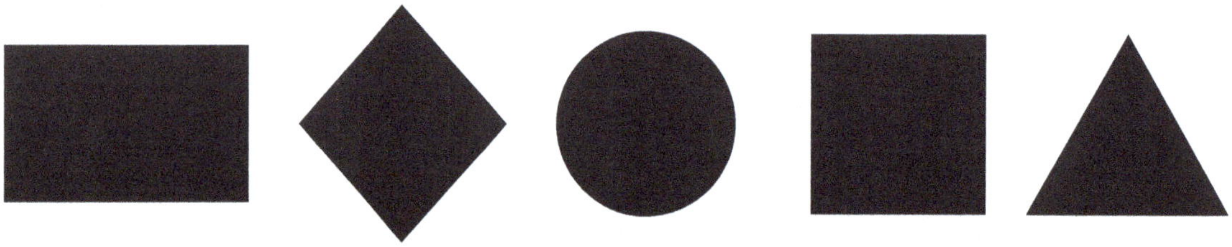

Let's Go on a Shape Hunt!

Take out 4 pieces of paper. Label the papers **circle**, **square**, **rectangle**, and **triangle**. Go around your house and draw pictures of as many things as you can find under each column. How many things can you find? Which shapes did you find the most of?

Color by addition.

3+3 2+1 3+2

2+2 1+1 6+1

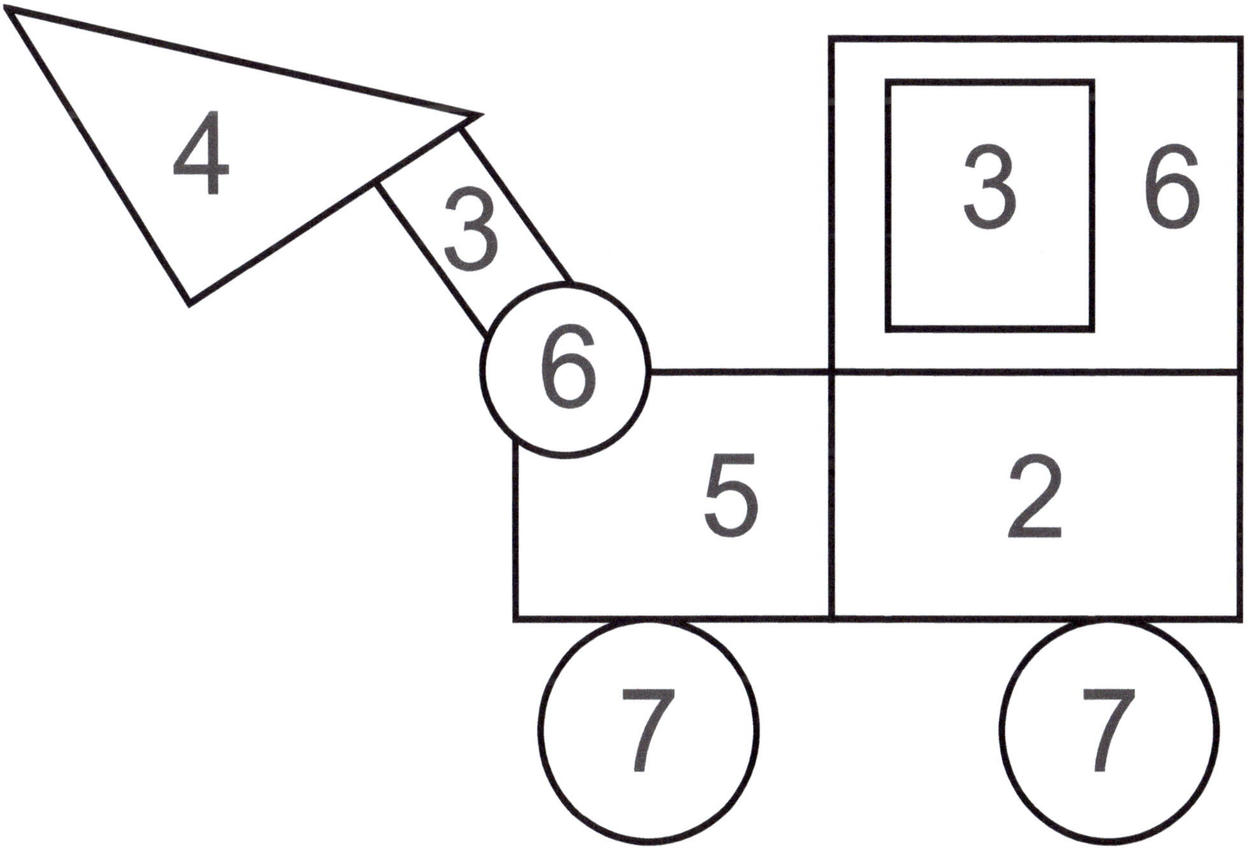

CHAPTER 4

Thinking about God's World

LESSON 31

Thinking about Differences

God, help us to know You so well that we will know how You want us to live. Amen.

Memory Work — *Practice counting from 1 to 10 forward and backward.*

> "My sheep hear My voice, and I know them, and they follow Me." (John 10:27)

Sheep will follow the shepherd because they spend time with him. He protects and cares for them. We can spend time with God by studying the Bible and talking to Him in prayer.

Can you find 5 differences in these pictures? Study the first picture and then compare it to the second picture. Circle the things that are different.

Find 5 differences in the ocean pictures.

Circle the things that are different. Color the pictures.

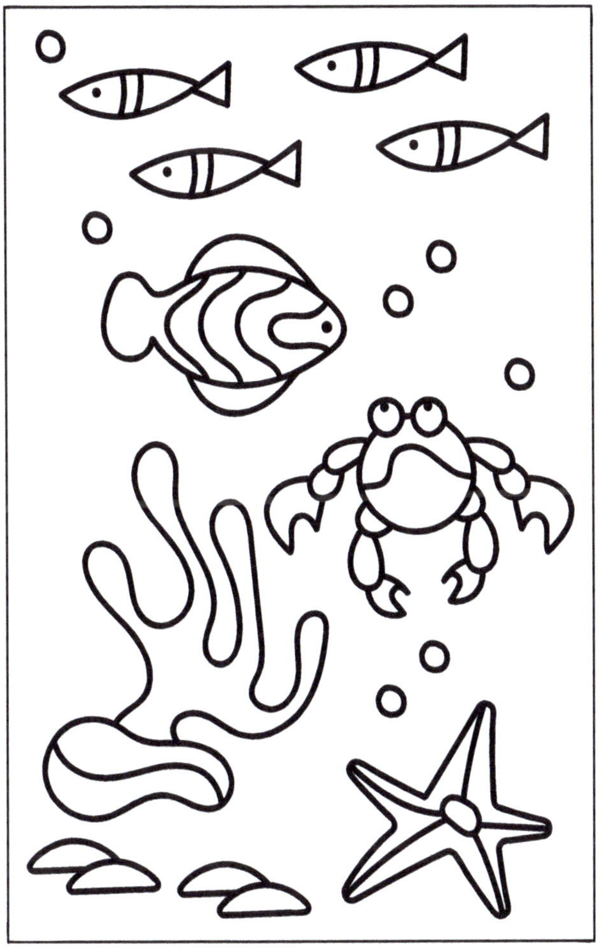

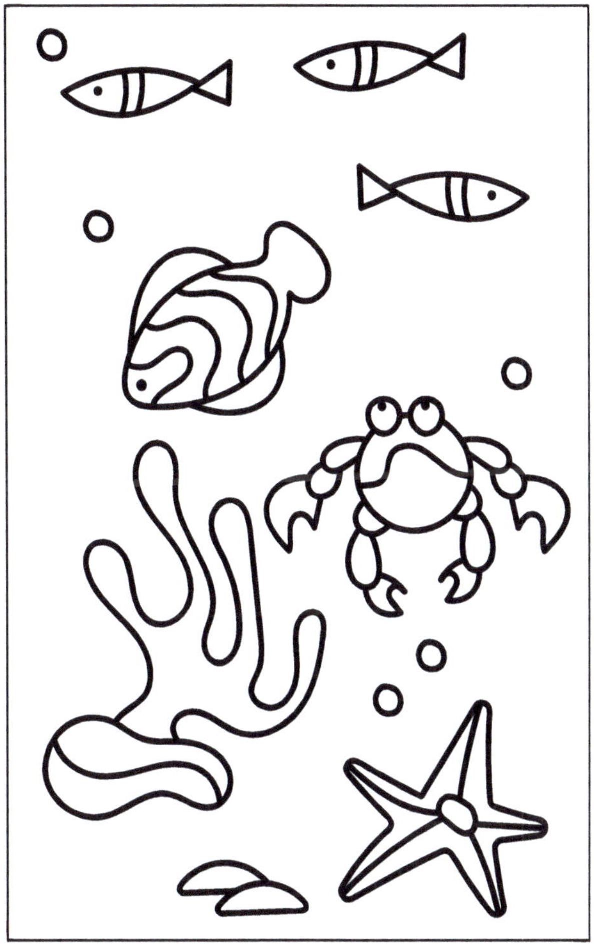

Let's Organize!

Ask your parents if there is a drawer or cabinet in the house that you can organize. Maybe you can arrange the things from big to small or by making groups of the same kind of things. If you arrange things neatly, it will be easy to find something when you need it!

LESSON 32

Addition Practice

God, thank You for the tools You give to help us learn math.

Memory Work — *Practice recognizing primary and secondary colors.*

Add these blocks by bringing them together.

Count the blocks in each group, and write the number you counted in the box.

☐ + ☐ = ☐

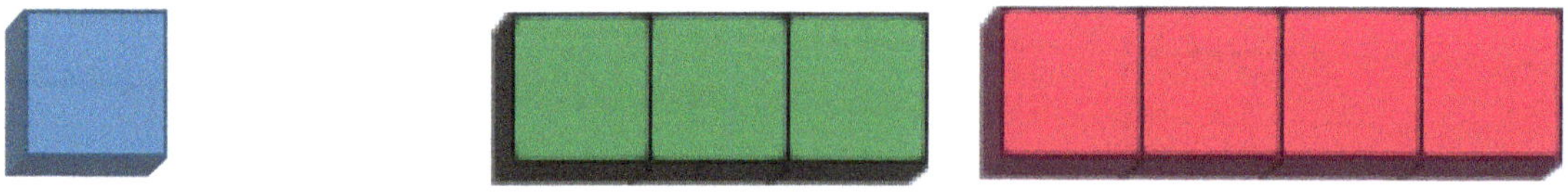

☐ + ☐ = ☐

☐ + ☐ = ☐

Draw in the missing blocks for these exercises:

LESSON 33 Critical Thinking with Shadows

God, help us to walk in the light with You each day. Amen.

Memory Work — *Practice drawing circles, squares, rectangles, and triangles.*

Shadows are dark areas made where something is blocking the light.

Can you match the animal to its shadow? Write the animal's number in the circle by its matching shadow.

Family Game Time!

Play a game of shadow tag. Access game #7 for directions on how to play the game: **Generations.org/kmath**.

People can have shadows too. Draw a line from the child to the matching shadow.

Time to Imagine!

Color the picture of the boy watering the flower. In the box next to it, draw what you think the shadows of the boy and the flower will look like.

LESSON 34 Thinking about Sizes

God, help us to do things in the order that pleases You. Amen.

Memory Work — *Practice guessing which of two objects your parent shows you is heavier, just by looking at it!*

Put on a pair of your parent's shoes and walk around the room. Do the shoes fit you? Think about how these shoes are different from your shoes. Think about the ways they are the same. When you think about things together this way, you are comparing them. Compare your shoe with your parent's shoe. Can you tell which one is bigger and which one is smaller?

Sometimes it is easy to compare. Sometimes you will need to look more closely.

Now find some old crayons. Try putting the crayons in order from smallest to biggest.

Number these bugs by size from 1 (smallest) to 4 (largest).
Color bug number 1 red, number 2 yellow, number 3 green, and number 4 orange.

God made big animals and small animals, and God made you. Draw a picture of yourself in the last box. Then number each of these from 1 (smallest) to 4 (biggest).

Circle the biggest picture in each row.

Watch and Listen!

Watch video #8 about comparing sizes at **Generations.org/kmath**.

LESSON 35

Critical Thinking with Sequencing

God, draw us closer to You one step at a time. Amen.

Memory Work — *Practice guessing which of two objects your parent shows you is heavier, just by looking at it!*

We get from one place to another by taking small steps, one at a time. The same is true about math. We learn little pieces day after day and year after year. Soon we will be answering bigger and harder questions.

Let's try to put things in order. Below, figure out what came first and what steps it took to finish. The snowboarder jumped over a ramp. Number the pictures in order from 1 (start) to 7 (finish).

Number the pictures of the chick hatching out of an egg.

Number and color the pictures of how the child grows.

Listen and Discuss!

As a parent reads the following statements, think about the order in which all of these happened, from 1 to 7. Can you write those numbers in the boxes?

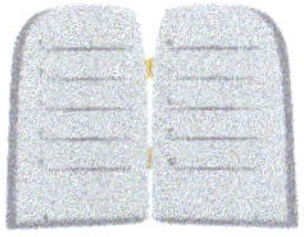
God gave Moses the Ten Commandments.

God made man and woman.

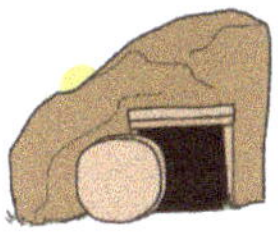
Jesus rose from the grave.

Noah built an ark.

Jesus was born as a baby.

God made the earth.

Jesus died on the cross.

LESSON 36 Making Bigger Numbers from 6 to 10

God, thank You for allowing us to be part of Your church. Amen.

Memory Work — *Practice recognizing primary and secondary colors.*

> And believers were increasingly added to the Lord, multitudes of both men and women. (Acts 5:14)

Today we will add bigger numbers together, to get an even bigger number!

When people believe in Jesus, they become part of God's church. God wants to use you to help His church grow. You can do this by inviting people to church and telling others what Jesus has done for you.

Let's add somebody to the church. When we add a person to a group, we need to bring them all together. We will start with 6 people, and then we will add 1 person to them. Let's bring them together like this!

6 + 1 = 7

Bring these popsicle sticks together like this! We start with 5, and add 3. Bring them together, and how many are there altogether? Count them!

5 + 3 = 8

Now, let's add these bigger numbers.

Add the blocks by bringing them together. Count the blocks in each group, and write the number you counted into each box. We have filled in some of the numbers for you.

☐ + 3 = ☐

☐ + ☐ = ☐

☐ + ☐ = 9

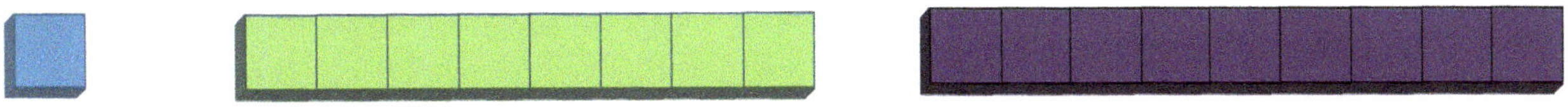

☐ + ☐ = ☐

Add It!

This family adds a child. How many people are there in this family now?

4 + 1 = ☐

This family adds twins to the family. When you bring these children into the family, how many people are there now?

4 + 2 = ☐

Fill in the missing numbers!

Watch and Listen!

Count the flowers in video #9 to practice your addition at **Generations.org/kmath**.

LESSON 37

Practice Making Bigger Numbers from 6 to 10

God, we praise You for our food and for the food You give the animals! Amen.

Memory Work — *Practice counting from 1 to 10 and drawing circles, squares, rectangles, and triangles.*

God provides food for the animals. Help the penguin find his fish by adding numbers along the path. You may use your manipulatives to help you count.

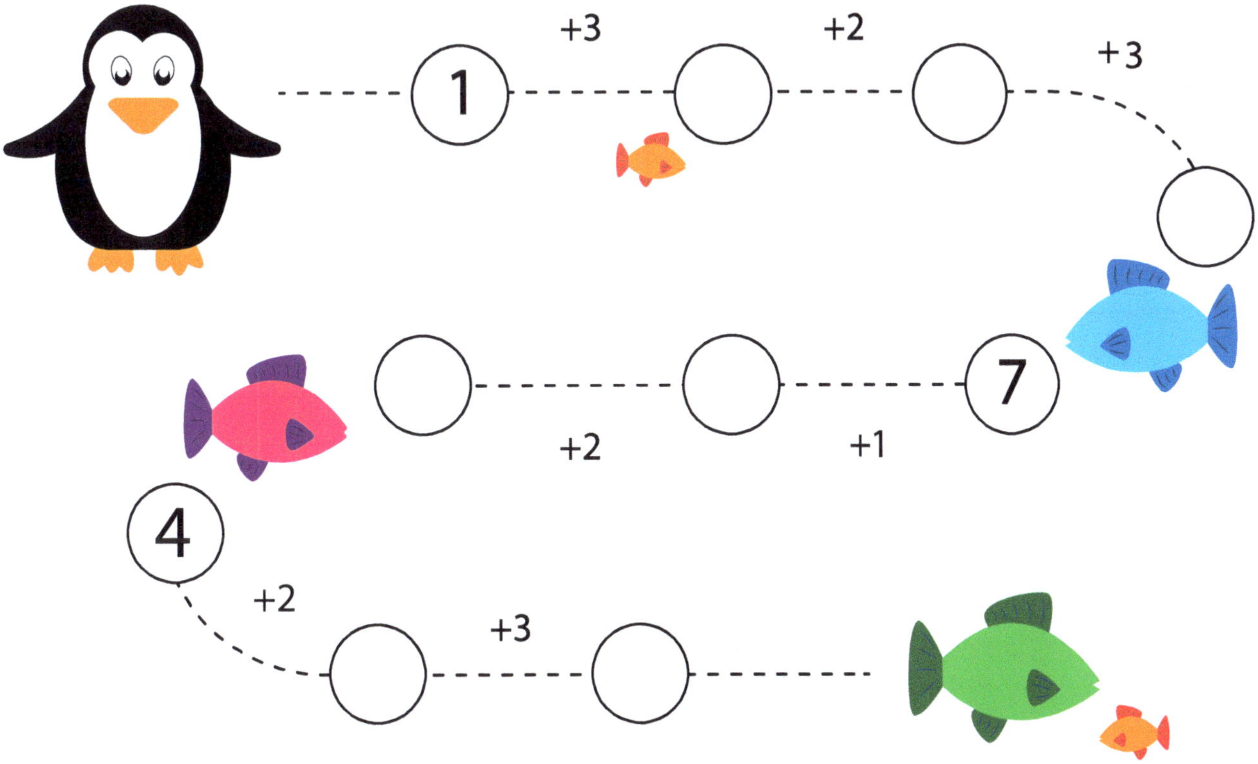

Play Pretend!

Let's pretend you are an animal finding its food. Choose a path that you will walk along at home. Have your parent roll 2 dice. Add the numbers that are rolled. If you are correct, take a step forward and pretend to eat your food. (You may need to roll again or ask for help if you roll a 5 and a 6 or two 6s.)

Find 7 differences between the pictures. Circle and count them.

LESSON 38

More Practice Making Bigger Numbers from 6 to 10

God, thank You for animal life! You have added so much beauty to the world we see. Amen.

Memory Work — *Practice counting from 1 to 10 forward and backward and knowing which numbers are bigger or smaller.*

When we add, we make bigger groups. Today we will make bigger groups of animals. Some animals like wolves, birds, and horses travel together in groups.

How many horses are joining the herd? What is the size of the herd now?

 +  = ☐

How many birds are joining the flock? What is the size of the flock now?

 +  = ☐

How many wolves are joining the pack? What is the size of the pack now?

 +  = ☐

How many chicks are joining the flock? What is the size of the flock now?

 +  =

How many cats in total are there?

 +  =

Practice counting as you connect the dots. Color the turtle.

LESSON 39

Even More Practice Making Bigger Numbers from 6 to 10

God, thank You that math can be fun. Help us to enjoy every chance we get to learn and play with our family. Amen.

Memory Work — *Practice counting from 1 to 10 and recognizing primary and secondary colors.*

Practice this addition. Use your blocks to help you get the right answer.

$3 + 5 = \square$	$2 + 6 = \square$
$0 + 8 = \square$	$6 + 3 = \square$
$1 + 3 = \square$	$2 + 3 = \square$
$4 + 5 = \square$	$10 + 0 = \square$
$7 + 3 = \square$	$3 + 4 = \square$

Play Beach Ball Toss!

Get a beach ball. Have your parent use a permanent marker to write addition questions with answers up to 10 all around the beach ball. Toss the ball back and forth. When you catch the ball, look at the question closest to your hand. Say the answer and throw the ball back again.

Sing It!

Here are some websites to help you practice adding by counting.

- Access video #10 to learn about adding numbers to 10.
- Access video #11 to learn about adding by counting.
- Access video #12 for adding up animals like penguins.

Math Game

Find the number that would make each additon problem correct. You may need a parent to help.

CHAPTER 5

Giving Money and Time

LESSON 40

Let's Count Bigger Numbers!

God, thank You for creating so many things for us to count!

Memory Work — *Practice counting numbers 1 to 20.*

> LORD, how many are Your works!
> In wisdom You have made them all;
> The earth is full of Your possessions.
> There is the sea, great and broad,
> In which are swarms without number,
> Animals both small and great. (Psalm 104:24-25 NASB)

How many fish do you see in this school?

How many ants are there on that anthill?

How many stars are there in the sky? More than ten!

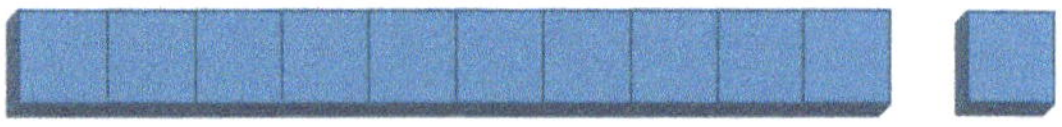

God has made bigger numbers for us so we can count big groups of things—like fish in the sea, ants on the ground, and stars in the sky.

So far you have learned to count to 10. Today you will learn to count bigger numbers! One way to picture big numbers is to group them in tens. That makes it a faster way to count. Look at the first number "11" below. Do you see a group of 10 blocks, and 1 more block next to it? That's 11! Can you say "eleven"?

Now let's say these numbers together!

11 **Eleven**

12 **Twelve**

13 **Thirteen**

14 **Fourteen**

15 **Fifteen**

16 **Sixteen**

17 **Seventeen**

18 **Eighteen**

19 **Nineteen**

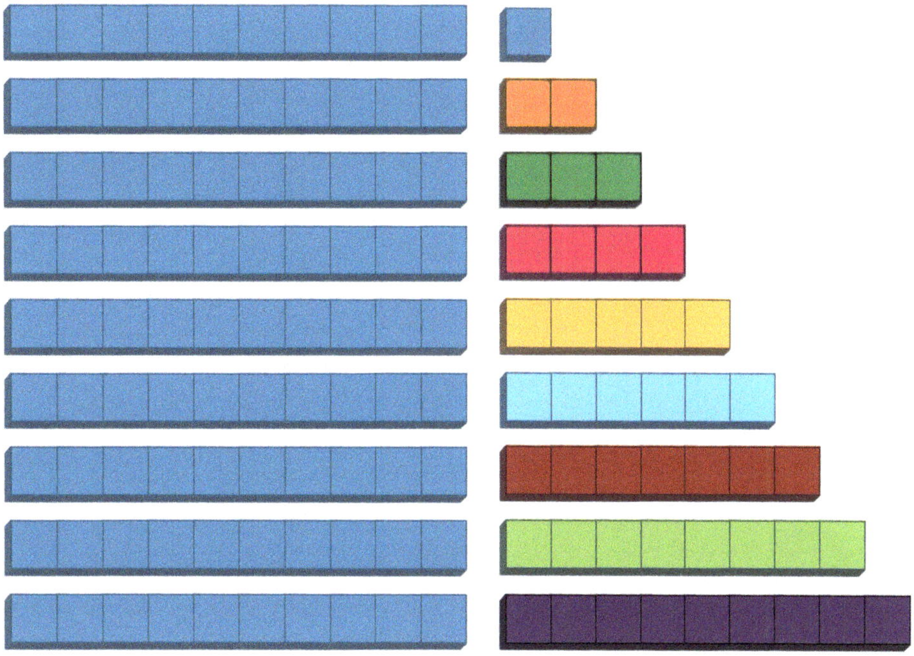

Trace the numbers from 10 through 19.

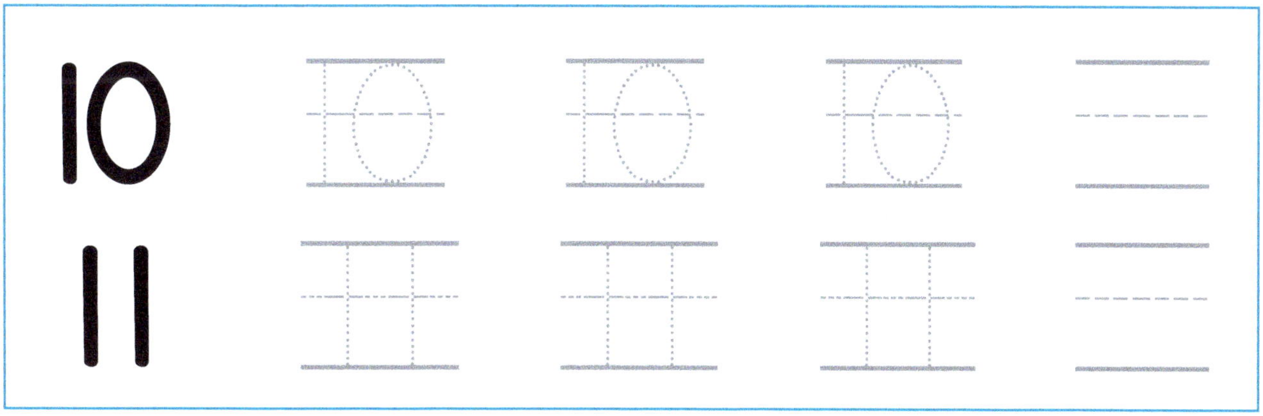

12

13

14

15

16

17

18

19

LESSON 41 Counting Even Bigger Numbers

Thank You, God, for making this world with so many things to count! Amen.

Memory Work — *Continue to practice counting from 1 to 20.*

> Moses told the people of Israel: "Your fathers went down to Egypt with seventy persons, and now the LORD your God has made you as the stars of heaven in multitude." (Deuteronomy 10:22)

Look at all those chairs! How many chairs are there in this picture? How many people will sit in these chairs? Count the rows. How many rows of chairs are there? That's right! Five rows. There are five rows of ten chairs.

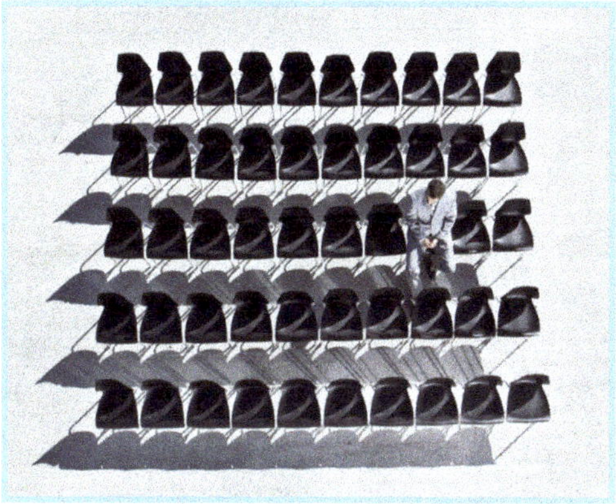

Today, we will learn to count numbers with more tens. After 19 comes 20! **Twenty**! That's two tens. Twenty! Two tens!

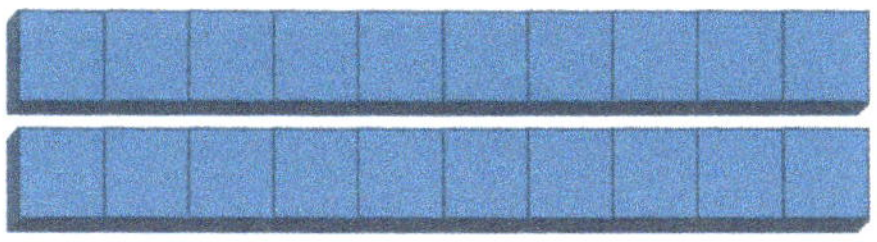

What is the number for 3 tens? **Thirty**!

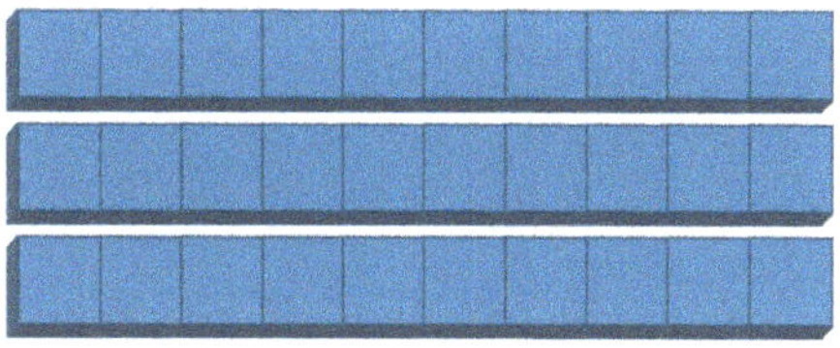

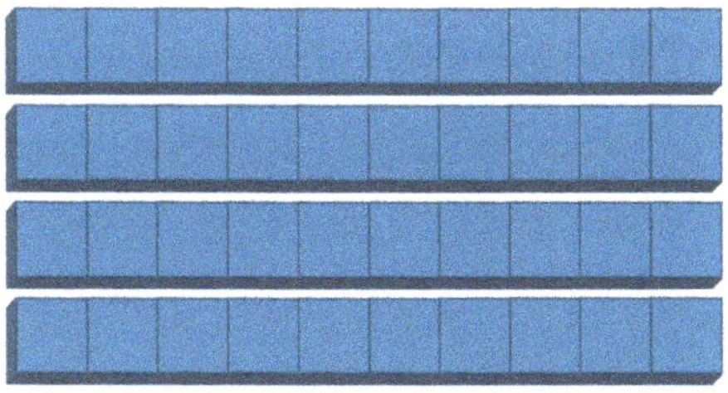

What is the number for 4 tens? **Forty**!

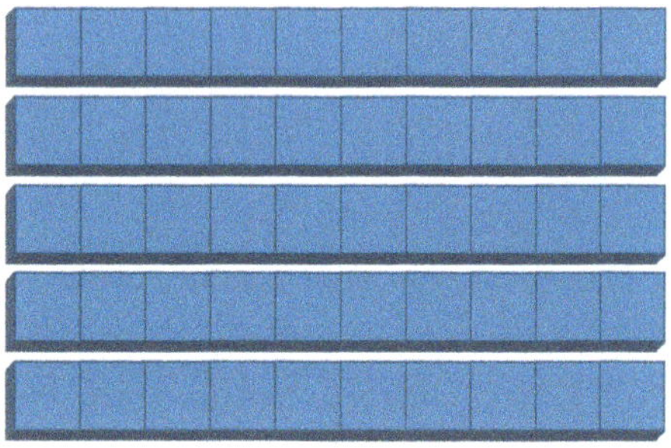

What is the number for 5 tens? **Fifty**!

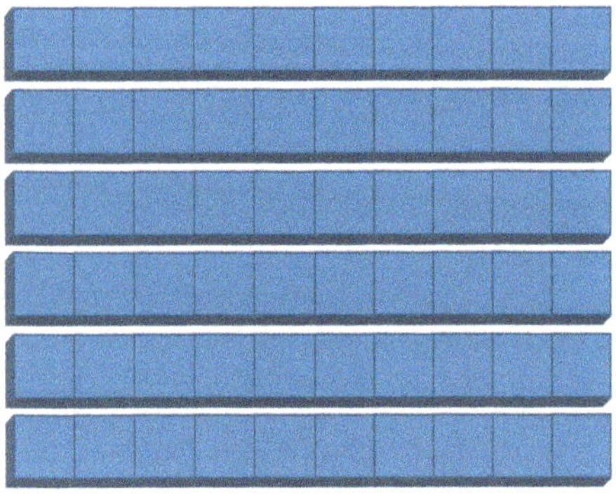

What is the number for 6 tens? **Sixty**!

What is the number for 7 tens? **Seventy**! That is the number of people in Jacob's family when they went live in Egypt so many years ago.

You have learned to count from one to twenty. Let's finish this lesson by counting from 20 to 29. Go ahead and practice writing numbers 20 through 29.

20 20 20 20

21 21 21 21

22	22	22	22	
23	23	23	23	
24	24	24	24	
25	25	25	25	
26	26	26	26	
27	27	27	27	
28	28	28	28	
29	29	29	29	

LESSON 42

What is Money?

God, You have given so much to us! Help us to give happily to those around us. Amen.

Memory Work — *Use your flash cards to practice your addition facts.*

...God loves a cheerful giver. (2 Corinthians 9:7)

We earn money for the work we do. Money is used to pay for things we need. The ability to work comes from God. We should remember to honor God by giving back to Him. When we give to God's church, it pays for things that the church needs. God is pleased when we cheerfully give.

In Bible times, money was made of gold and silver. These coins are what money looked like:

Today, money is mostly made out of paper. Money may look different depending upon where you live. Here is what money looks like in the United States.

This is one dollar. This money is made of paper. We write the amount of money with a dollar symbol in the front like this: $1.

This is five dollars. See the number 5 in the corners. It is written like this: $5.

This is ten dollars. The number 10 is in the corners. It is written like this: $10.

This is twenty dollars. The number 20 is in the corners. It is written like this: $20.

Which is More?

Circle the money that is more than the other for each of these. Write the number for each amount in the box provided.

Which of the coins in the rows below is missing on this treasure island?

Practice It!

Use the cutout pictures of money that come with this book. Practice telling a parent what each piece is called. Do you have real money of your own? If so, see if you can tell your parent the name of each piece.

Watch It!

In video #13, you will see the differences between the coins. You will also learn who the famous people are on the front of American coins at **Generations.org/kmath**.

Parent/Teacher: References to "solving the equation" are the same as finding the answer, or adding to find a sum.

LESSON 43

Adding up Money

God, all things come from You! Help us to use our money wisely. Amen.

Memory Work — *Practice pointing out primary and secondary colors around you.*

Do you like to go shopping with your family? We do not want to be greedy with our money, or greedy for more things. But God gives us money to buy things we need like food and clothing.

Today we will practice counting our money. You may want to use real money to do this.

Suppose you have \$5 and three \$1 bills. How much do you have altogether? When counting your money, start with the bill with the biggest number and add the rest of the smaller bills. A \$5 bill and three \$1 bills are added this way:

$$5 + 3 = 8$$

(Count starting with 5 + 1 [= 6] + 1 [= 7] + 1 [= 8])

Try adding these up yourself!

Add by counting the dollars.

+

=

+

=

+

=

+

=

It's good to save our money for a time when we need it.

Practice adding the money you have saved in your piggy bank.

3 + 4 = _____	6 + 2 = _____
1 + 5 = _____	4 + 2 = _____
1 + 5 = _____	4 + 2 = _____
7 + 3 = _____	2 + 5 = _____
3 + 2 = _____	1 + 8 = _____

Let's Give Some Away!

Color the picture. While you color, talk to your parent. Imagine you have saved up a lot of money. Who are some of the people you could help with your money? What would you want to give them? Did Jesus give to those in need?

LESSON 44

Giving Away our Money to Help Others

God, we worship You with our giving. Amen.

Memory Work — *Practice adding up to 10 with paper money.*

...but she out of her poverty put in all that she had... (Mark 12:44b)

God uses people like you every day to meet the needs of His people. He may want you to give to your church, to a Christian ministry, or to a person that you know.

Match coins with the number and the church offering basket that were given by the children. Color the picture.

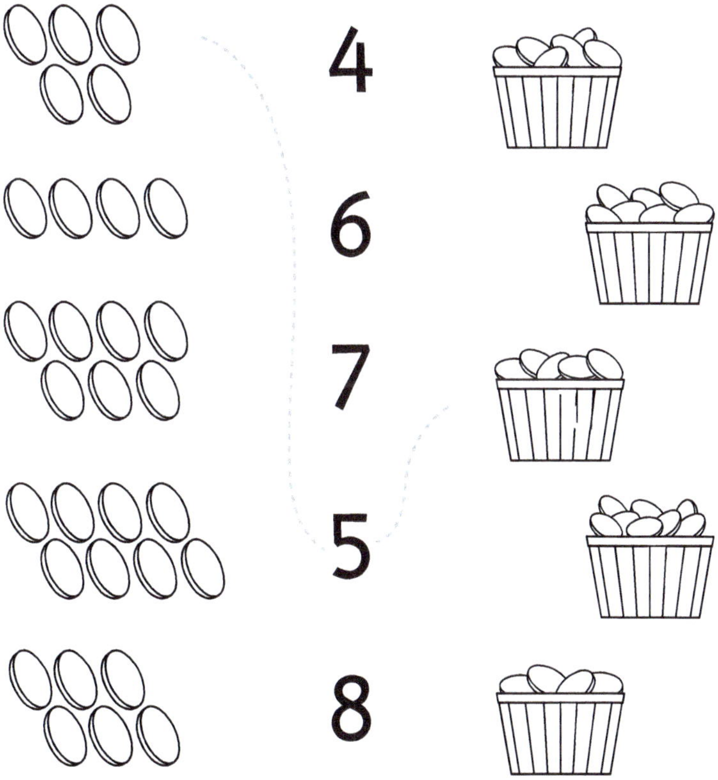

Watch It!

In video #14, see how the widow was willing to give all she had.

Parent/Teacher: video contains a cartoon image of Jesus.

Add the numbers on the piggy banks and match the coin with the answer.

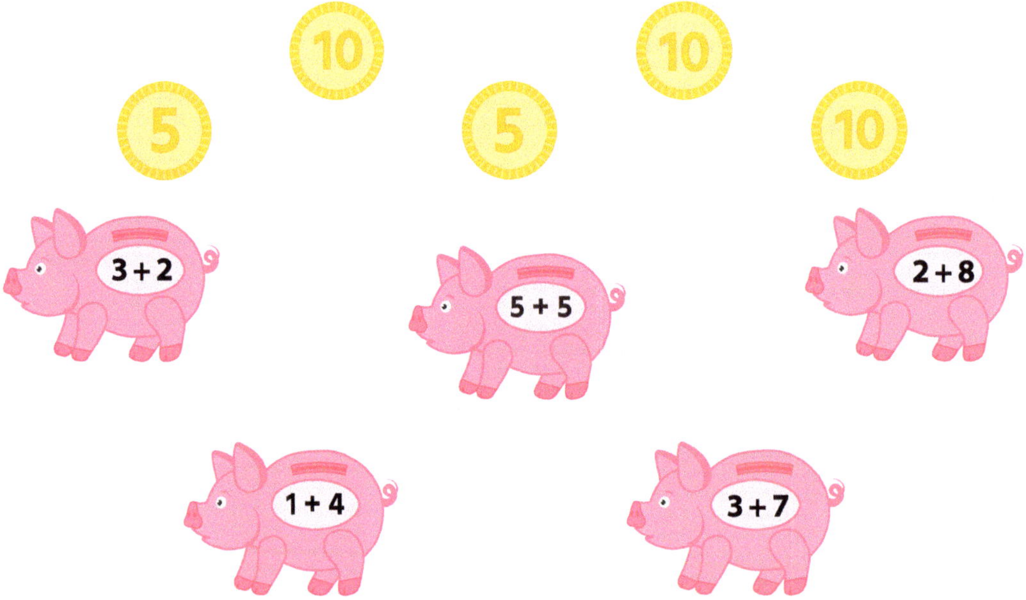

Earn Money!

Maybe you have heard the saying "money doesn't grow on trees." God provides work for people to do to earn money to pay for things.

Ask your parent if there are any chores around the house that you can do to earn money by working. Make sure to do your best at any job you do!

Let's Play Store!

Use the cutout money included in the book, and play store with your parent. Pretend you are buying some of your toys. Practice counting the money that you will need to buy the toys.

LESSON 45

Giving Our Time

God, we want to spend time with You. Help us to set aside time to talk to You in prayer often. Amen.

Memory Work — *Practice putting several books or toys in a line from lightest to heaviest.*

> Look carefully then how you walk, not as unwise but as wise, making the best use of the time, because the days are evil. (Ephesians 5:15-16 ESV)

God does not want us to waste time. He wants us to spend our time helping others. We spend time with those that we love. We show God that we love Him by praying and reading our Bible. Life is often busy, so we need to learn to use our time wisely. There are 24 hours in a day. Spend that time with God and family to show them you love them.

We measure time with a clock. Clocks have numbers that you can count from 1 to 12. There are two hands on a clock that spin. One points to the hour and the other points to the minute. For these lessons, you will learn about the hour hand.

Trace and color the clock. Have your parent show you where to write the numbers in the correct spaces.

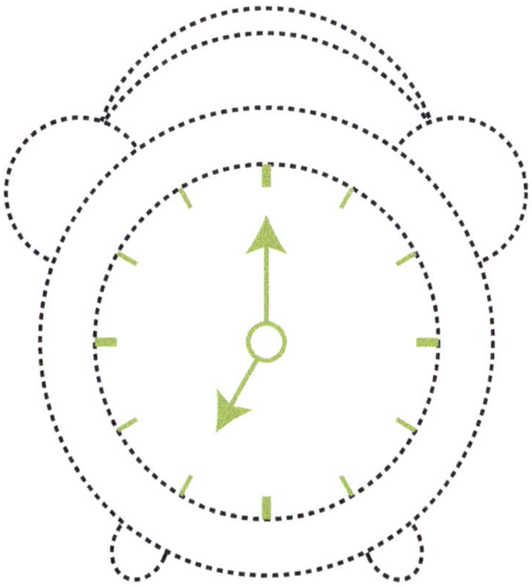

Tell Time!

When the long minute hand is pointing straight up, we use the word "o'clock." Do you see the short hand in the first picture? It's called the hour hand, or little hand. Look at these clocks. Where is the little hand pointing? If the little hand is pointing at 1, then you say, "One o'clock!" Let's do this together!

Watch It!

Learn to read time on a clock in video #15 at **Generations.org/kmath**.

Extra Challenge

Students may advance to the second video if they have mastered the first video.

LESSON 46

Practice Telling Time

God, help us to remember to praise You throughout the day!

Memory Work — *Practice addition using your flash cards.*

Where is the little hand (the hour hand) pointing? Now you are telling the time! Write the number next the word "o'clock." The first one is done for you.

o'clock

o'clock

o'clock

o'clock

o'clock

o'clock

o'clock

o'clock

What does the clock say?

With a little help from a parent, let's draw in the hands for. . .

When you wake up in the morning:

When you have breakfast:

When you have lunch:

When you have dinner:

When you go to bed at night:

LESSON 47

Using Clocks

God, guide our day so that we will learn to plan ahead. Amen.

Memory Work — *Practice adding sums of 10 with your block cutouts.*

What time do you begin your day? Do you follow the same routine each day? Planning ahead can help us to use our time wisely. Once the important tasks are done, we can have time to play.

Talk to your parent about what time you do some of the activities shown in the picture and write the hour in the box.

Make a Human Clock

Take a piece of chalk and draw a large clock on the ground somewhere outside. Lie inside the circle and use your arms to point to the times your parent says.

Play It!

Access game #8 for a time-telling game at **Generations.org/kmath**.

LESSON 48

Practice Math with God's Money and Time

God, help us to use the gifts You give for Your glory.

Memory Work — *Practice drawing and labeling circle clocks.*

How much money do you see here? Write the number in the box. Circle the biggest number.

Write the number of blocks in the box.

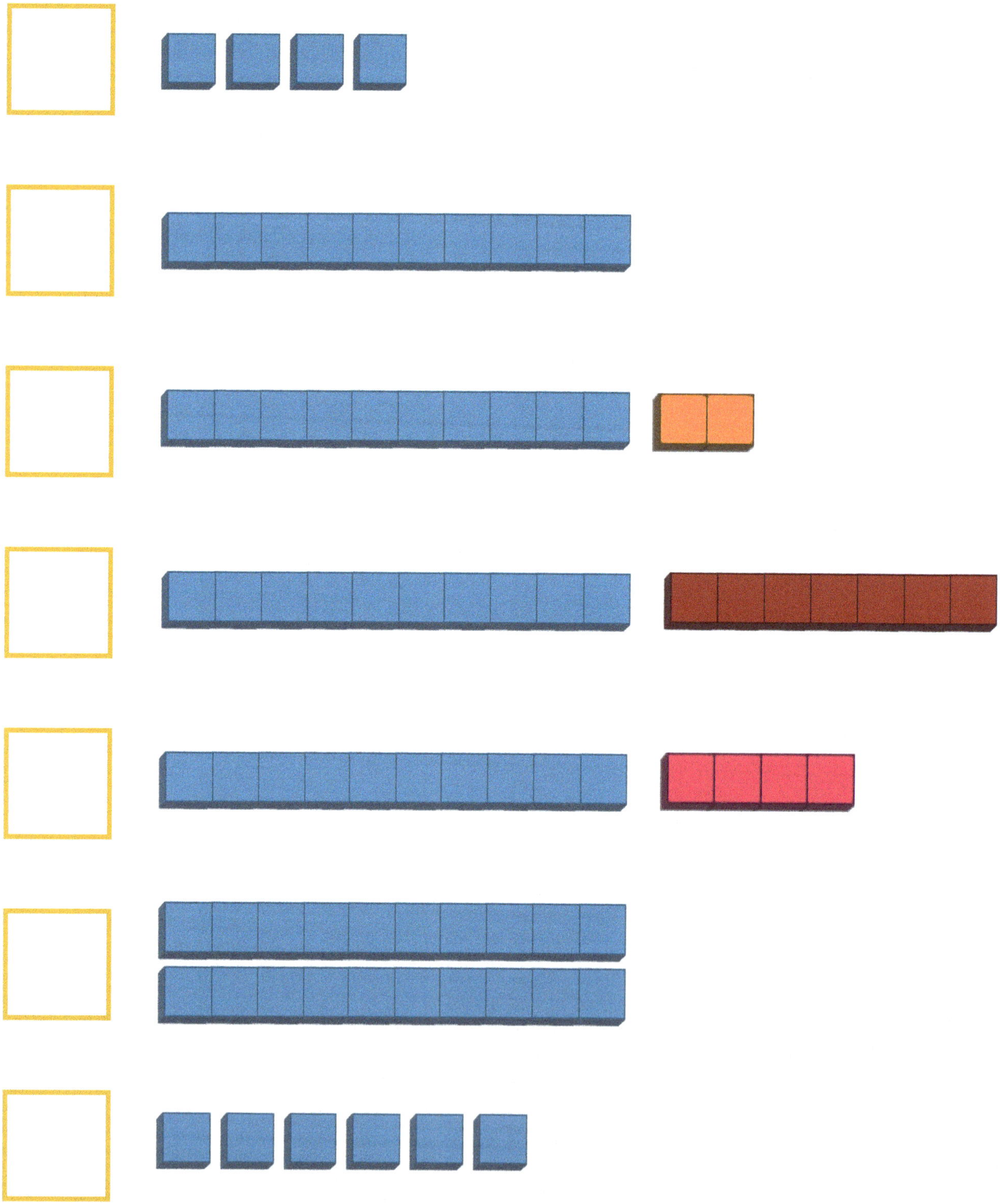

Make It!

Cut out the clock and hands. (You may need your parent's help.) Use a pushpin or fastener to allow the hands to spin. Practice using the clock hands to tell time like the clocks at the bottom.

THIS PAGE IS LEFT INTENTIONALLY BLANK

What time is it? Write the number the hour hand is pointing to in the box.

☐ o`clock

☐ o`clock

☐ o`clock

☐ o`clock

☐ o`clock

☐ o`clock

CHAPTER 6

Measuring God's Creation

LESSON 49

How Big Is That?

God, You are so big, and we are so small. Help us to remember this today! Amen.

Memory Work — *Practice counting from 1 to 25.*

> Now concerning the pillars: the height of one pillar was eighteen cubits [27 feet]. (Jeremiah 52:21a)

God has made big things and little things. We make big things and small things too. How tall are those pillars? How tall are you? God made you to grow bigger each year. Maybe your parents measure your height sometimes. Thank God that you are growing bigger each year. We measure things to see how big or small they are. We use tools like rulers to measure things.

Let's measure something today! Some people measure the height of little boys and girls using inches. Others measure the height of little boys and girls using centimeters. Rulers are usually made to measure both ways. One side measures in centimeters. The other side measures in inches. Let's measure this carrot in centimeters. First, we line up one side of the carrot with the 0. Then, we look at where the carrot ends. Write the number that you see. God grew the carrot to this length!

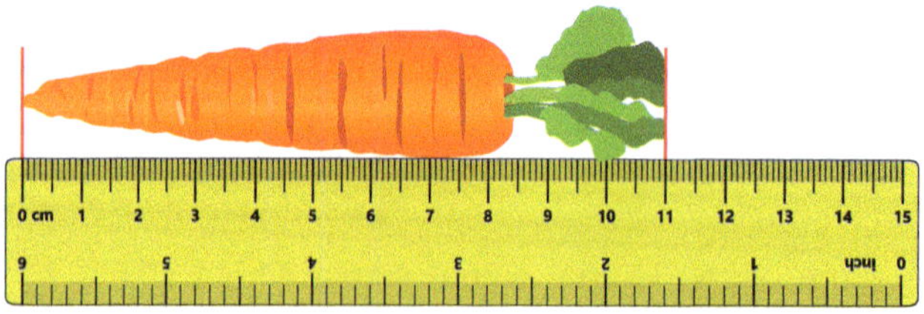

_____ cm

Now try measuring these items.

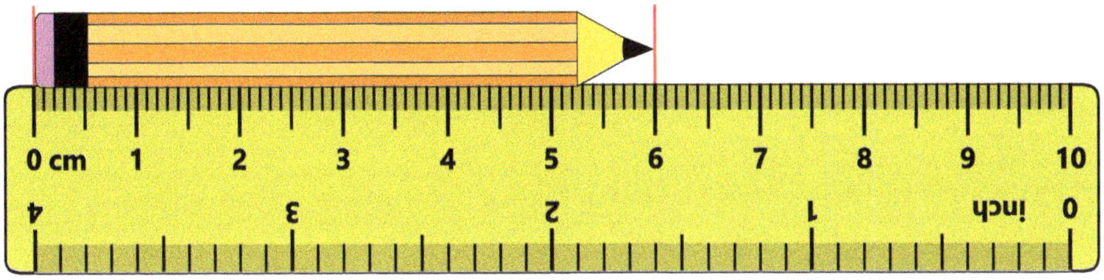

___ cm

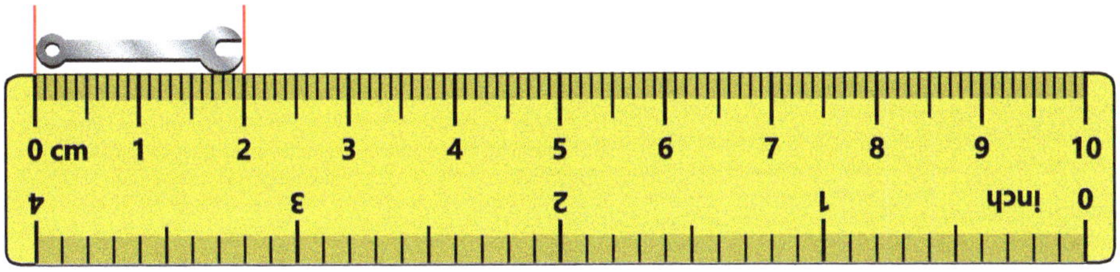

___ cm

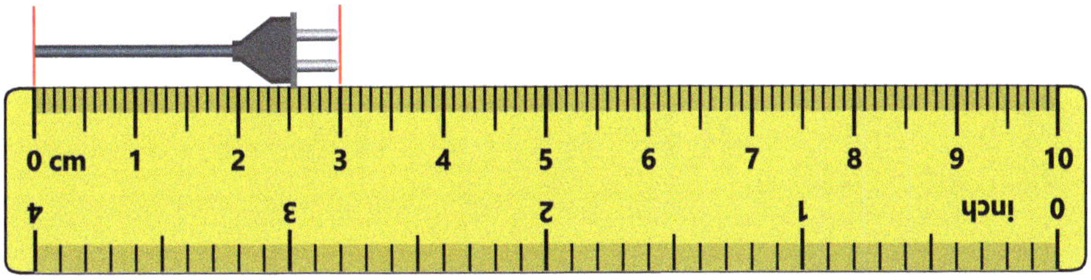

___ cm

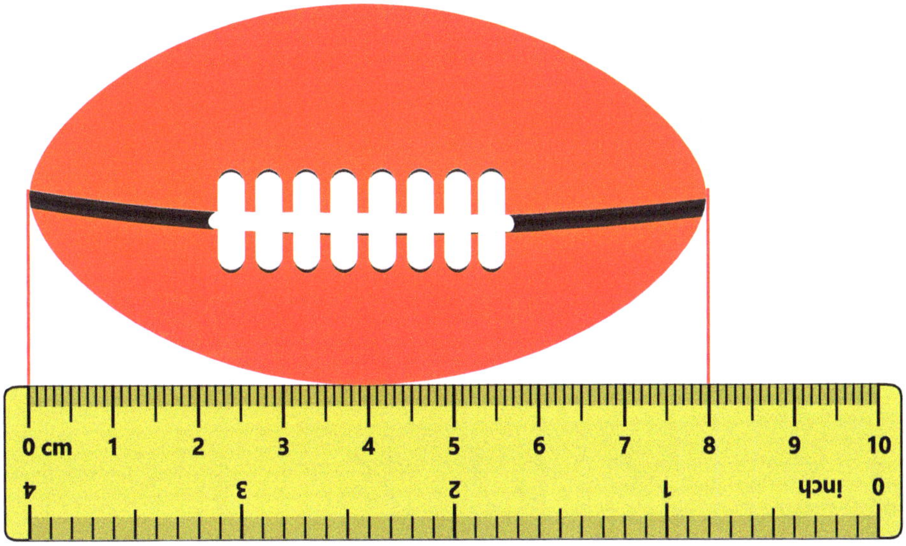

___ cm

Do you have a ruler at home? See if you can measure some of the things in your house.

LESSON 50 Practice Counting and Addition

God, thank You for helping us learn many ways we can use your numbers!

Memory Work — *Practice telling time and counting money.*

Can you fill in the missing numbers?

1	2	3	☐	5	6	7	
7	8	☐	10	11	12	13	
4	5	6	☐	8	9	10	
13	14	15	☐	17	18	19	
20	21	22	23	24	☐	26	
23	24	25	26	27	28	☐	

Add these numbers together! Put the two groups of bananas together and count them to make sure you have the right answer.

$1 + 1 = \square$

$2 + 2 = \square$

$2 + 3 = \square$

$4 + 1 = \square$

Match the pictures with the right shapes and colors. Write the number in the circle.

LESSON 51

Measure More Lengths

God, help us to praise You with every gift You have given us. All good gifts, big and small, have come from you!

Memory Work — *Practice naming primary and secondary colors.*

Measure the pencils. Write the length in the boxes below.

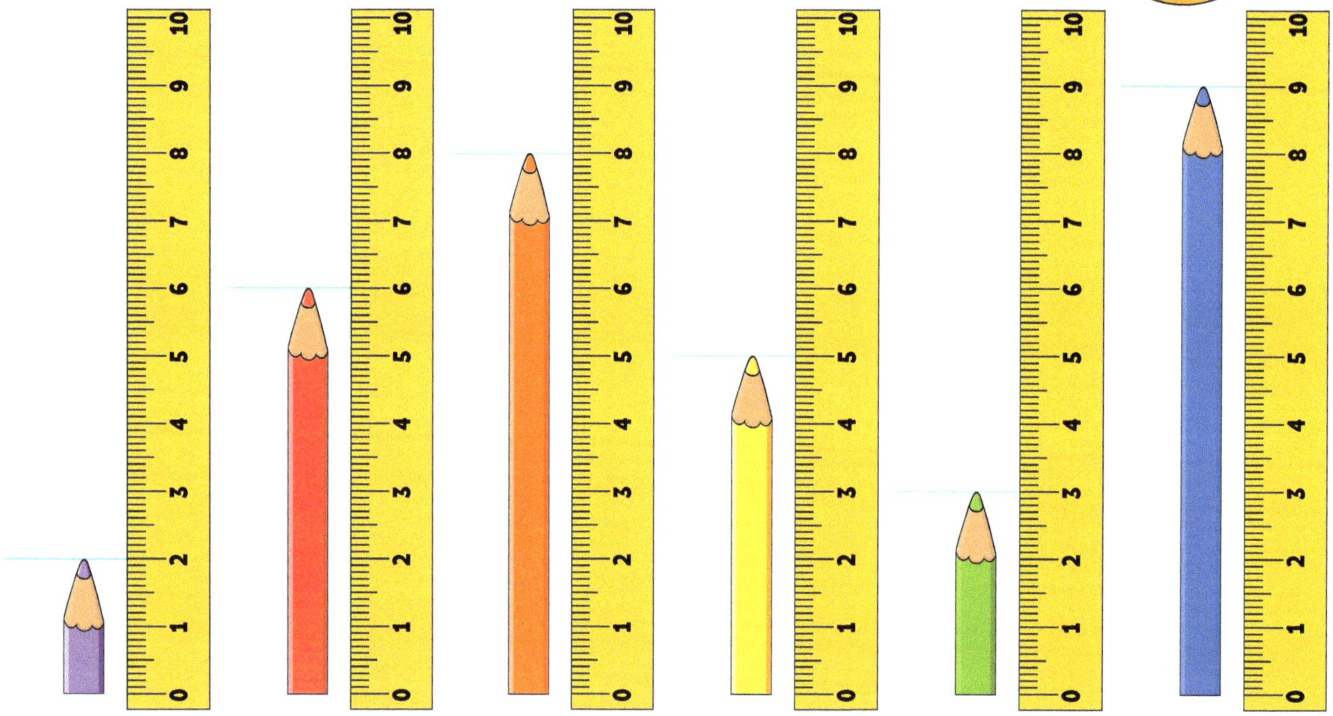

How tall are you?

If you have a tape measure, have an adult help measure you.

Now, put two pencils together. The little purple pencil is 2 cm. The bigger red pencil is 6 cm. Measure the length of these two pencils put together. This is the same as adding 2 + 6.

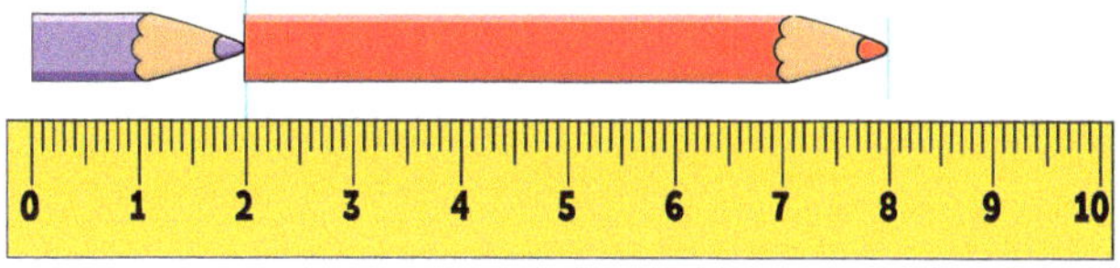

How long are these pencils put together like this?

Color the picture. Write the numbers on the measuring stick to indicate how many feet tall the child has grown.

Measuring helps us to build things like houses. Complete the dot-to-dot to create a house. Color your picture.

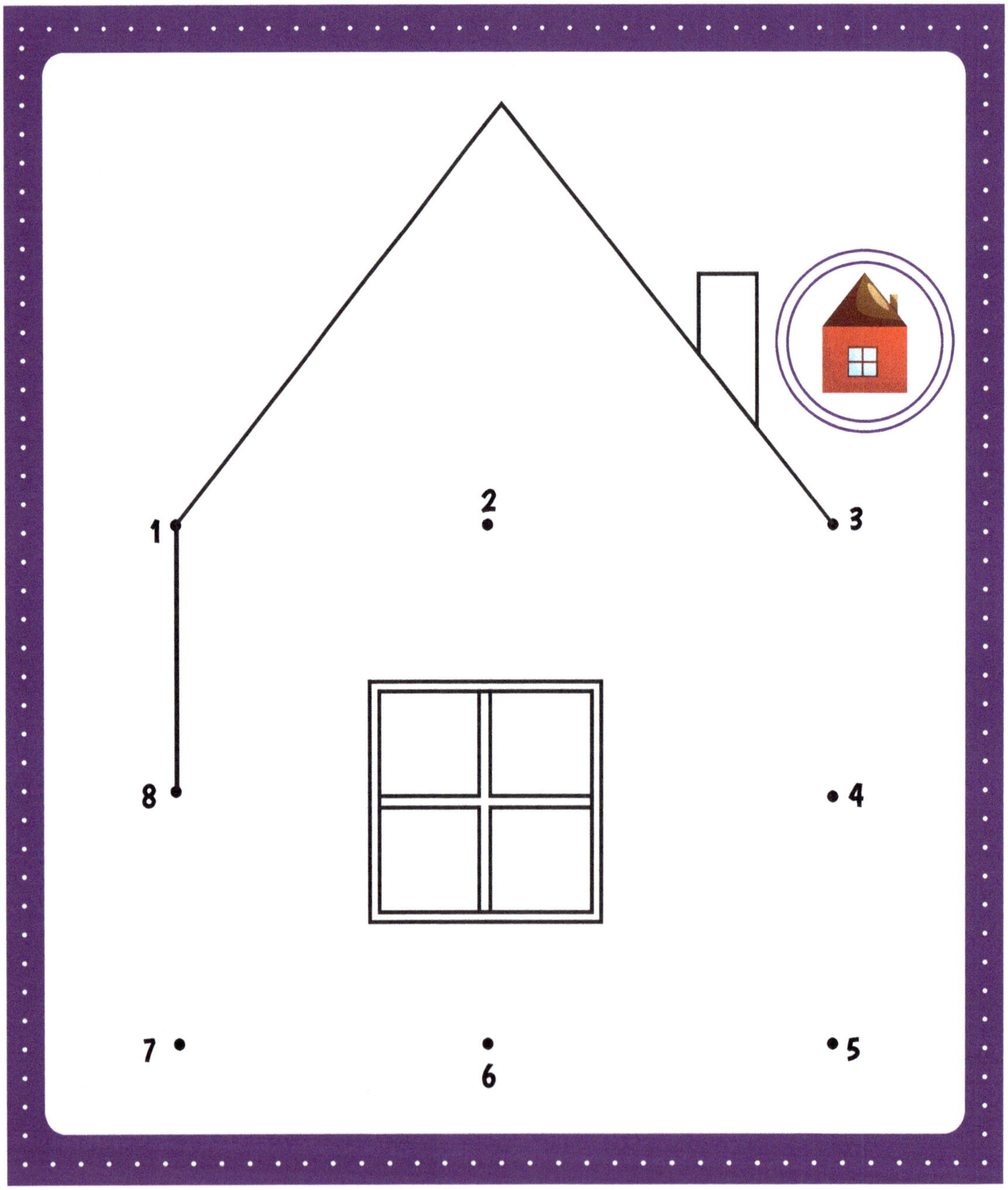

Sing It!

Listen to the measurement song: video #16 at **Generations.org/kmath**.

LESSON 52

Liquid Measurements

God, You are so big! Nothing we make could be bigger than You!

Memory Work — *Practice counting money.*

Now there were set there six waterpots of stone… (John 2:6a)

Do you have a favorite cup? We measure liquids using cups or bowls. We pour liquids into containers. If you don't put liquid into a container, it will spill all over the place. Liquids will take the shape of their containers. Some cups are bigger than others, and that means they can hold more liquid.

Americans mostly measure liquids in gallons and quarts. But other people measure liquids using liters.

It would take 4 quarts to fill up a gallon bottle.

And it would take a little less than 4 liters to fill up a gallon bottle.

That means a gallon is much bigger than a quart and a liter.

Gallon **Quart** **Pint** **Cup**

Match the bottles with the amount of liquid they hold. Each measurement will match two bottles.

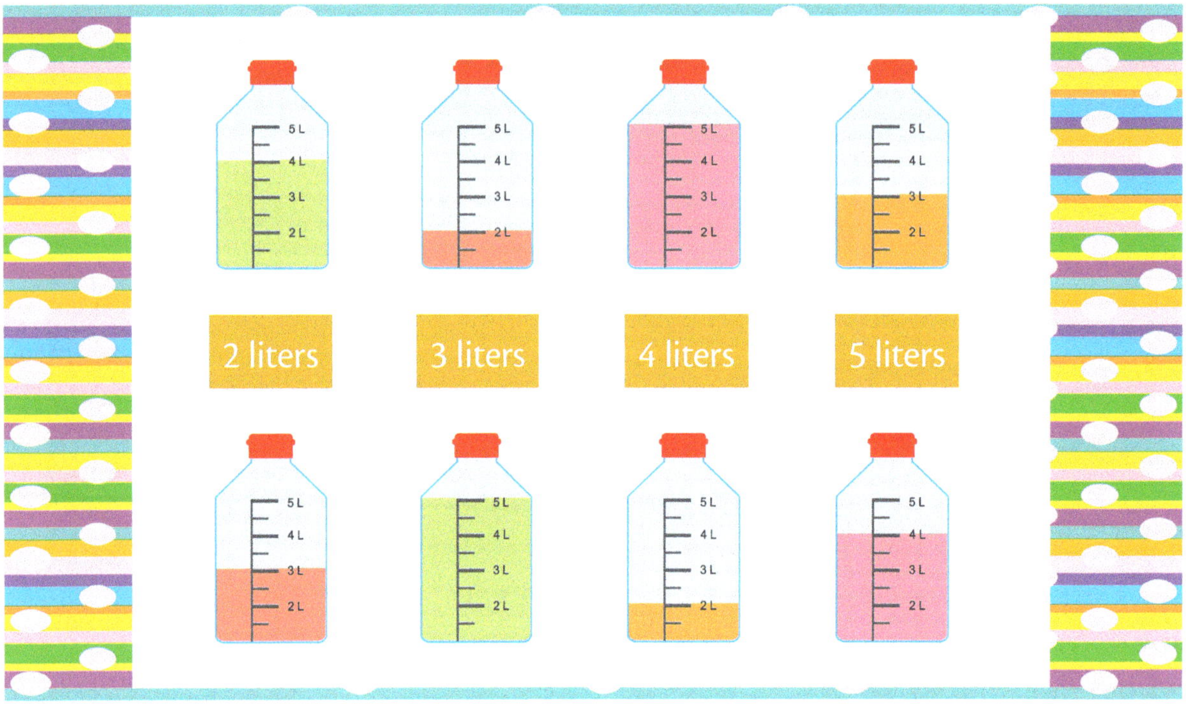

Watch It!

See how much different containers hold. View video #17 at **Generations.org/kmath**.

Experiment!

Now try to conduct your own experiments! Fill different cups or containers with water. Which ones hold more?

Bake It!

Ask your parent if you can help cook in the kitchen. You can measure other things besides liquids in cups and containers. You can measure sugar and flour like you measure milk, water, and oil. Remember to follow the recipe! And measure the ingredients carefully!

LESSON 53

Practice Measurements

God, thank You for all the water that You give us.

Memory Work — *Review addition facts to 10.*

Number the bottles of milk from lowest (1) to highest (5).

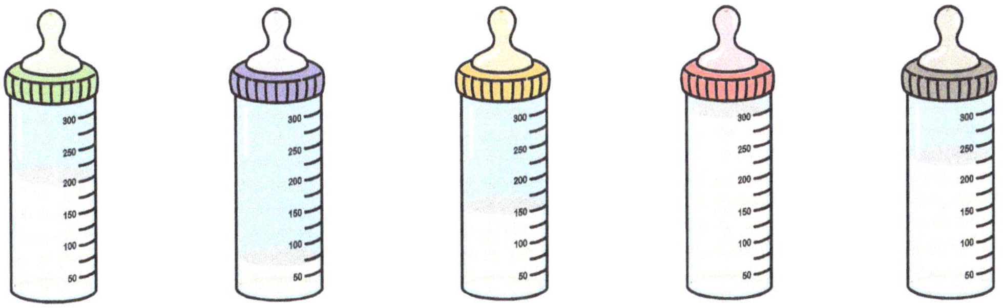

☐ ☐ ☐ ☐ ☐

For the next week, keep track of how many glasses of water you drink each day by coloring in one water drop for each glass.

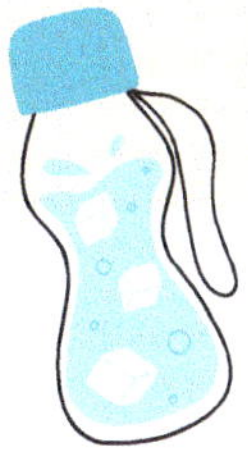

MON
TUE
WED
THU
FRI
SAT
SUN

Let's add these together! What is the total length of both pencils? Write the answers in the box!

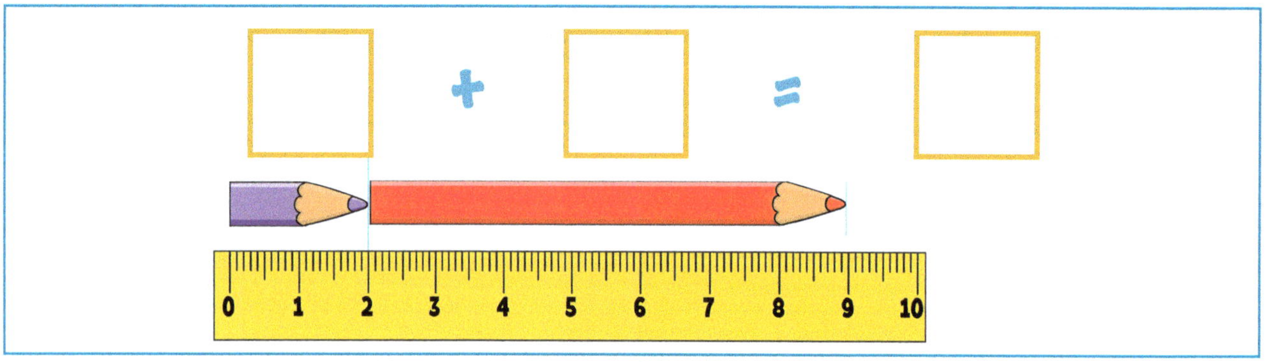

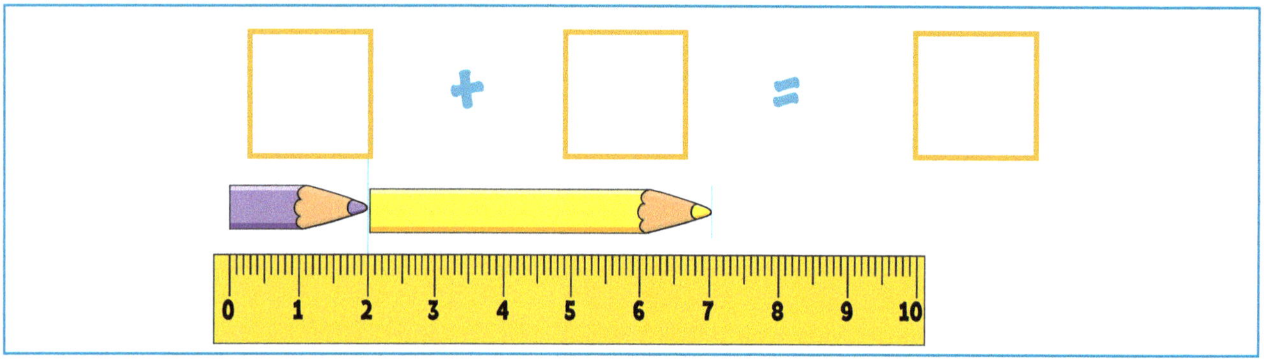

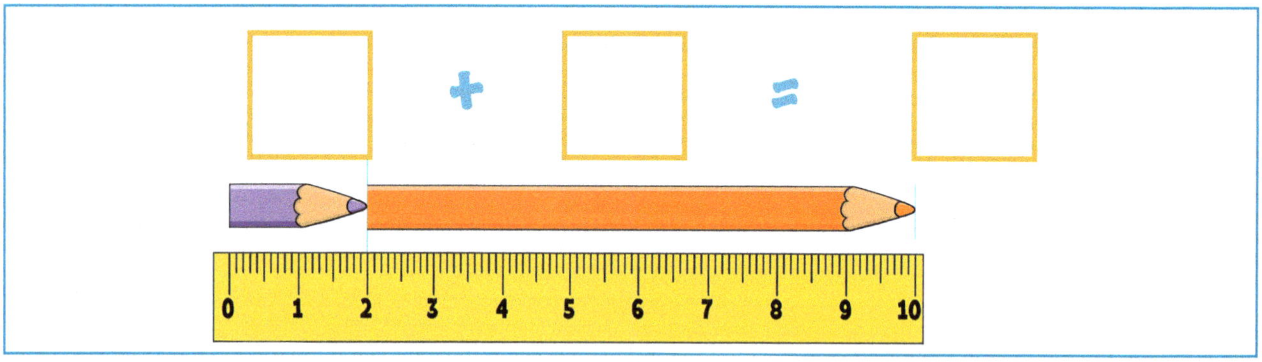

LESSON 54 Weight Measurements

Lord, You are strong! There is nothing you cannot do! Your strength amazes us! We can trust you to help us every day.

Memory Work — *Practice counting to 20 forward and backward. Practice measuring with a ruler.*

You have already learned how to measure lengths. How tall is that rhino? How long is that pencil? You have learned how to measure how much liquid is in a container. Here is another way to measure something: how heavy is it?

Thankfully, nothing is too heavy for God! He can lift everything. He can lift mountains, the sun, the moon, and the stars! We measure weight with a scale. Americans usually measure weight using pounds. Many other people measure weight using kilograms. Can you say, "kilogram"? A kilogram is heavier than a pound. A watermelon might weigh 11 pounds. But that's only about 5 kilograms.

When we set something on a scale like the one you see in the pictures on the next page, a little pointer will point to a number. That's the weight of the stuff on the scale. Write the number of kilograms the scale points to under the picture.

Write the number of kilograms the scale points to.

How much is this man lifting? Add the weights together in each of these rows.

+ 3 = ☐

+ 2 + 1 = ☐

+ 2 + 1 = ☐

+ 3 + 2 = ☐

+ 1 + 1 = ☐

How much do you weigh?

If you have a bathroom scale, measure yourself to find out how much you weigh.

LESSON
55

More Weight Measurements

God, lift us up when we fall!

Memory Work — *Look around your bedroom and find two big objects that are lighter than two smaller objects.*

Here is another kind of scale. We call it a balance scale. For this scale to balance, the weights of the things on both sides must be the same. If one side is heavier than the other, the lighter side will be higher, and the heavier side will be lower.

Circle the heavier side for each of the scales in the pictures. How many oranges will you have to add to the lighter side, so both sides will be the same?

$2 + \square = 4$

$3 + \square = 5$

$1 + \square = 2$

$4 + \square = 5$

Circle the heavier object on these scales. Draw an arrow downwards on the side of the scale that is lighter.

Watch It!

Watch video #18 about weighing vegetables. See **Generations.org/kmath**.

Try It!

If you have a kitchen scale, weigh different ingredients in your pantry. See which ones weigh more when you compare a variety of boxes, bags, cans, bottles, or jars.

LESSON 56

Distance

God, thank You for going the distance to rescue us!

Memory Work — *Practice drawing shapes and comparing bigger or smaller.*

"And whoever compels you to go one mile, go with him two." (Matthew 5:41)

At the beginning of this chapter, you learned about length, which is the measurement of how long something is. Distance is how far apart two things are from each other. Length and distance belong together. Let's review measuring first.

How long are these colored pencils? Write the number of centimeters in the box next to the ruler.

We measure the length using a tool like a ruler. The length shows us the distance from the back end of the pencil to the point.

Distance is kind of like length. But distance is the word used for measuring how far it is from one place to another. We usually measure large distances in miles or kilometers. A mile is a little bit longer than a kilometer.

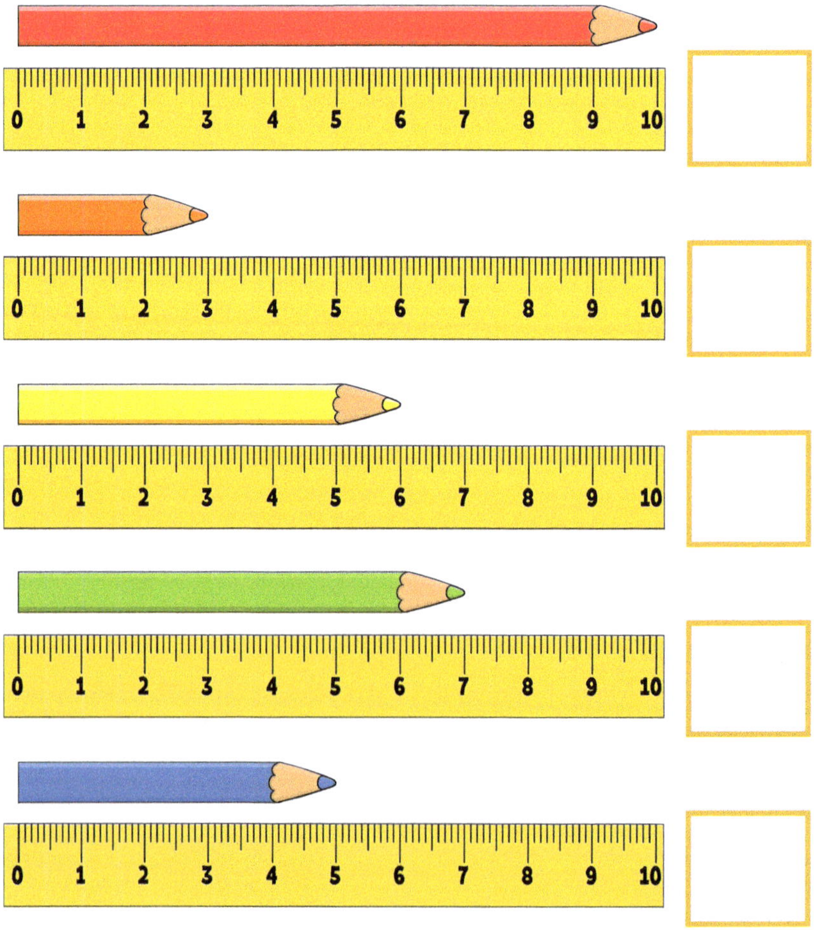

Ask your mother or father how far it is from your house to the following places. Write in the number of miles or kilometers. Then, circle the biggest number. This is the farthest place from your home!

Walk with God everywhere you go!

From your house to the nearest store

From your house to the library

From your house to church

From your house to your friends' house

From your house to your grandmother's house

Activity

You can measure distances by steps too!

- How many steps do you take to go from the kitchen to the nearest bathroom in your house?
- How many steps do you take to go from your bedroom to the kitchen?
- How many steps do you take to go from your bedroom to the front door of the house?
- How many steps do you take to go from the front door to the street?
- Color the box that shows the longest distance orange. Color the box that shows the shortest distance green.

Trace the path showing the distance the girl traveled on her scooter. Color the picture.

There are two ways to go from your house to the church on this map. Which is the shorter distance? Which is the longer distance?

The next time your family drives to church, ask your parent to track the real distance you travel.

LESSON 57

More Distance

God, no matter how far we go, You will be there!

Memory Work — *Practice guessing which of two objects in your kitchen weighs more before you pick them up.*

What is your favorite way to travel a far distance? Count each type of vehicle and write it in the circle below the picture. At the bottom, write your numbers again into the line and add up the total.

◯ + ◯ + ◯ + ◯ = ◯

A cat wants to travel a distance in his car. Cut out the pieces and glue them together onto a sheet of paper.

THIS PAGE IS LEFT INTENTIONALLY BLANK

Different vehicles travel distances at different speeds. Use different colors to color the path that each vehicle takes. Which vehicle is the fastest?

Solve the addition problems. Write each number in place of the pictures, then add.

 =3L  =2L  =1L

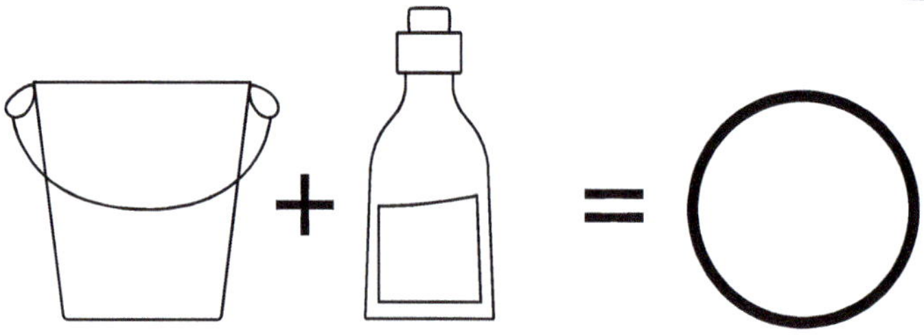

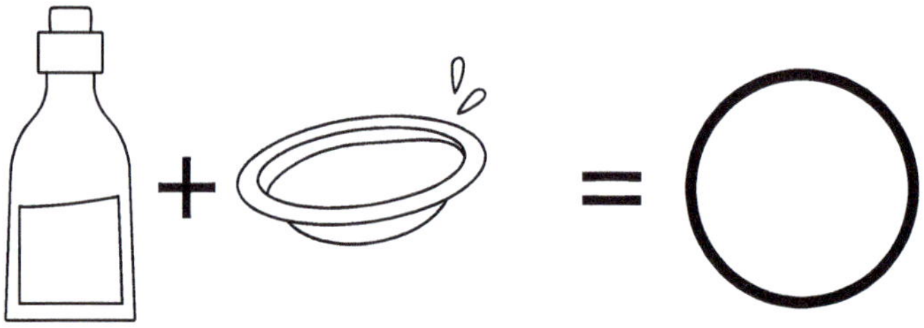

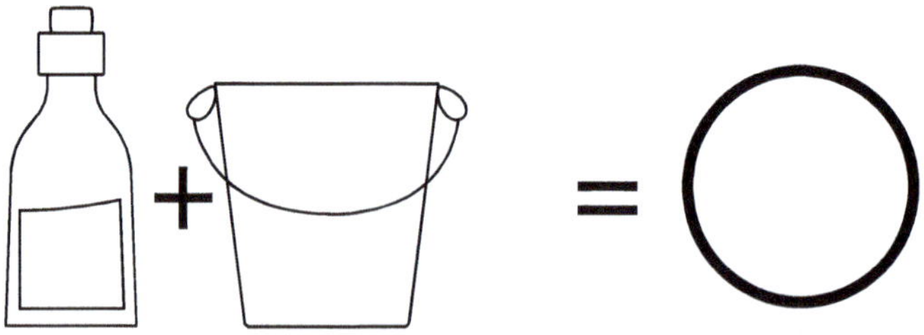

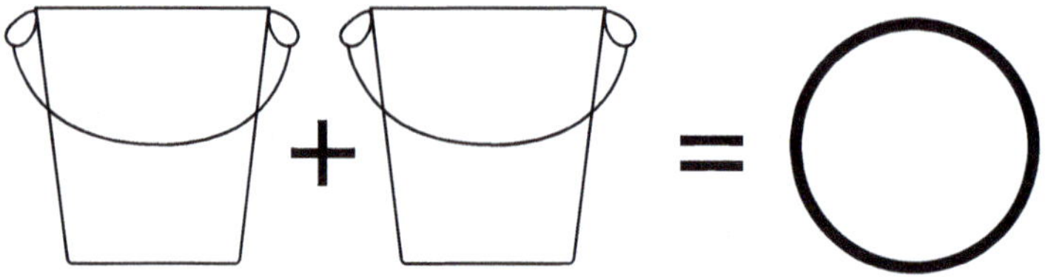

Time to praise Him!

Sing in worship to God along with this song. God is everywhere we go. Access video #19 at **Generations.org/kmath**.

CHAPTER 7

Let's Make Bigger and Smaller Numbers!

LESSON 58

Adding on the Number Line

Thank You, God, that we can share the good news about Jesus! Bring more sheep to the Shepherd!

Memory Work — *Practice counting from 1 to 25 and practice measuring with a ruler.*

> And the Lord added to the church daily those who were being saved. (Acts 2:47b)

We are like sheep, and Jesus gathers more sheep every day. He adds more of them to the church. We welcome new people to the church!

Today, we will add numbers using the number line. Let's add numbers by taking 2 jumps on the number line.

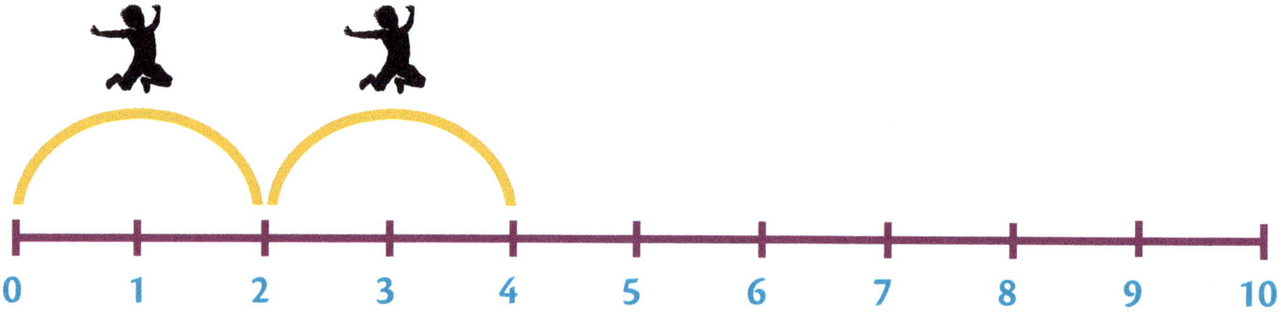

The first jump will be from 0 to 2. How many spaces did we jump? Now, let's take another jump of 2 more spaces. Where did we land? Write the number in the box.

2 + 2 = ☐

Let's take two different jumps! How far did we jump this time? Write the number in the box.

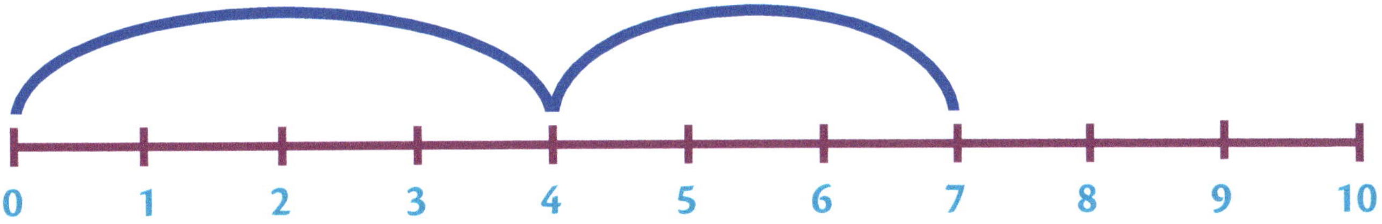

4 + 3 = ☐

Let's take two more jumps! How far did we jump this time? Write the number in the box!

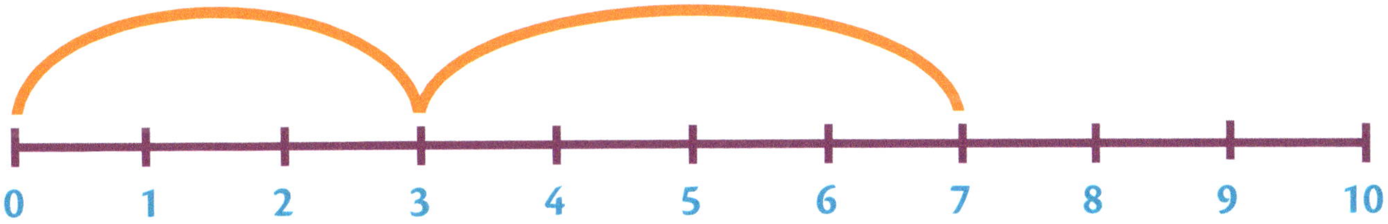

3 + 4 = ☐

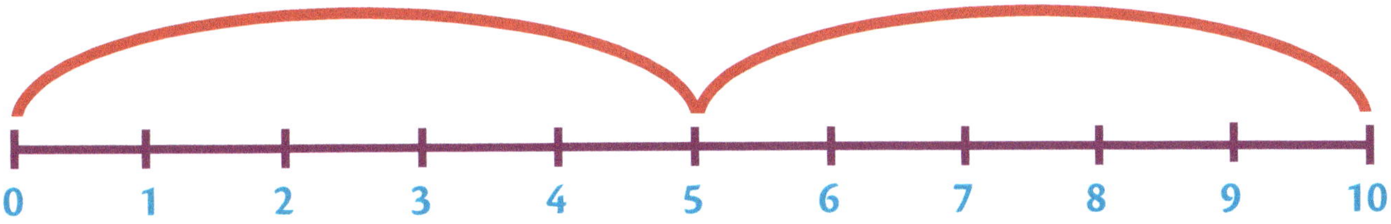

5 + ☐ = ☐

Watch It!

This will help you understand how to use a number line. Access video #20 at **Generations.org/kmath**.

Dot-to-dot

Count the numbers as you connect the dots. Color the picture.

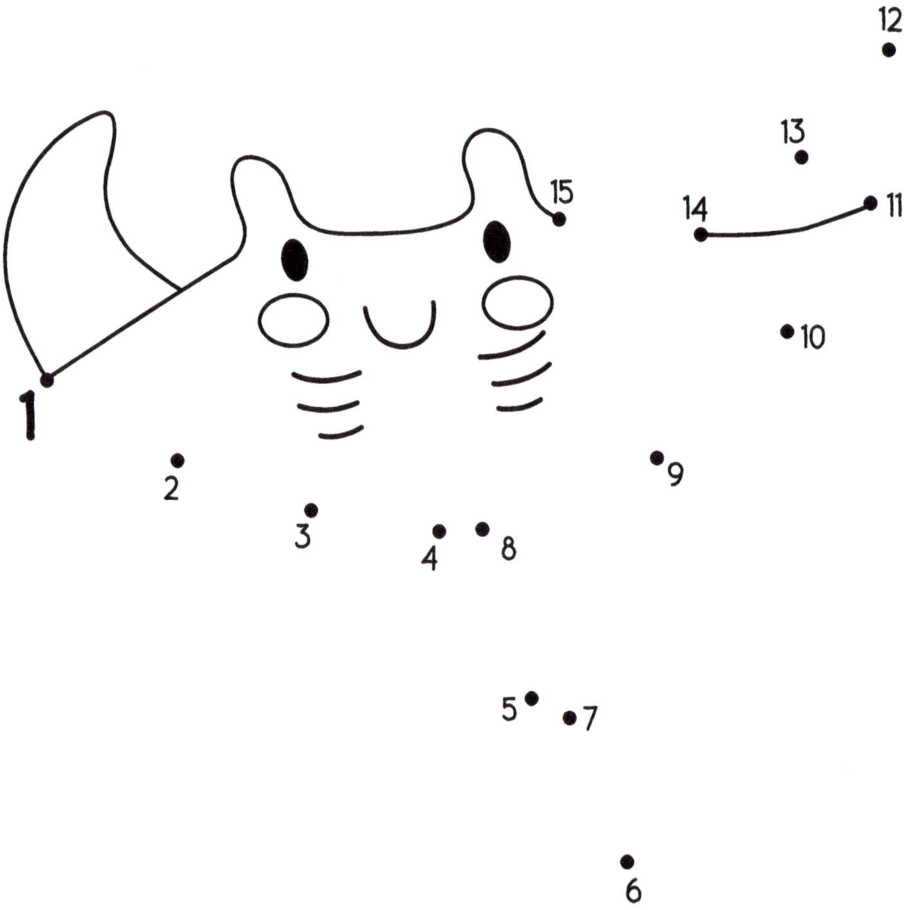

Count the emergency vehicles. Circle the correct number.

LESSON 59

Adding Bigger Numbers

O Father in heaven, thank You for the many blessings you have given to us. They are more than we can count!

Memory Work — *Practice addition facts with flash cards.*

Today we will add some bigger numbers on the number line.

Let's take two big jumps. How far did we jump this time? Write the number in the box.

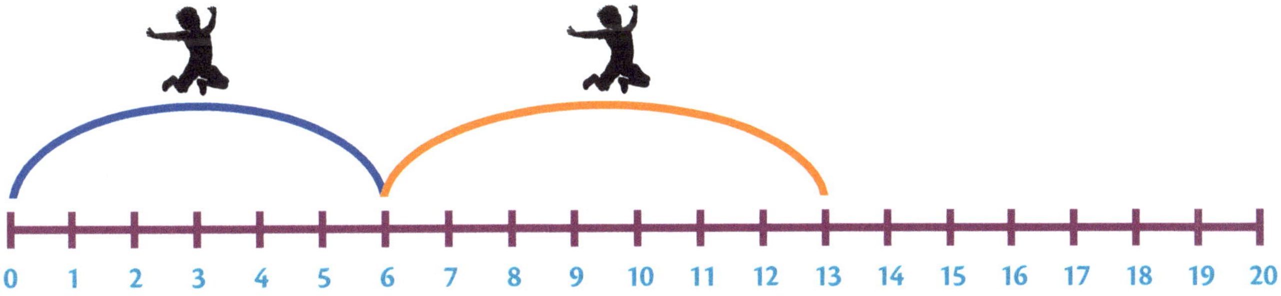

6 + 7 = ☐

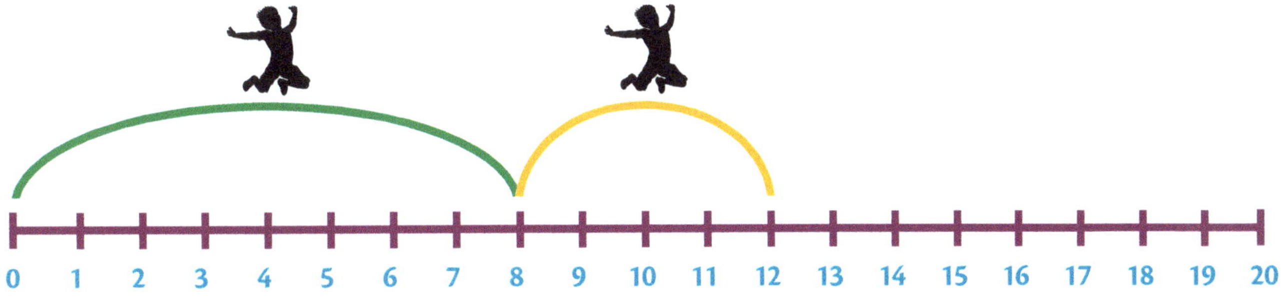

8 + 4 = ☐

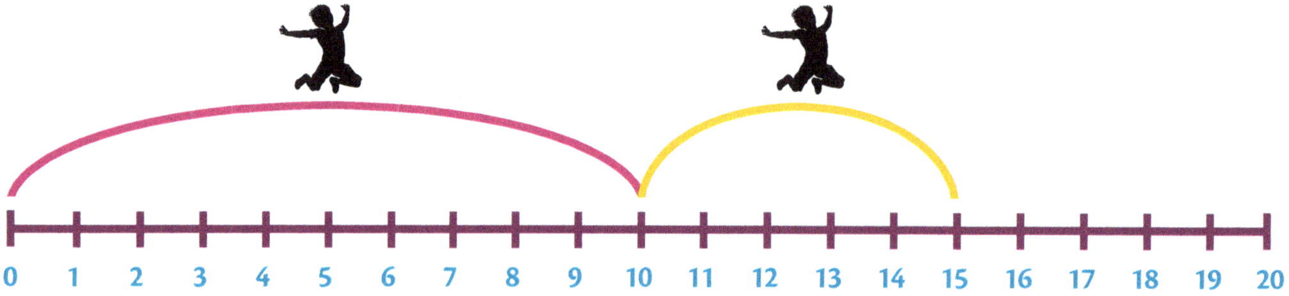

10 + 5 = □

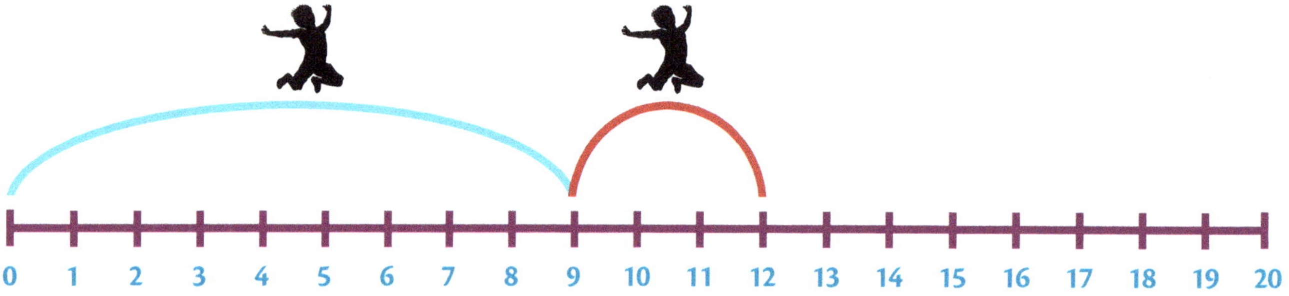

9 + 3 = □

Watch It!

We can praise God by counting and adding all the blessings He gives to us. Begin your day by listening to a song like "Count Your Blessings." Try to sing or play along. God loves to hear our praise. Access video #21 at **Generations.org/kmath**.

Add these bigger numbers together. Make two big jumps on the number line to get the answers. Write the number of colored dots into each box.

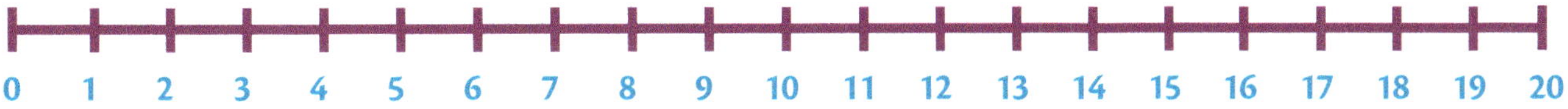

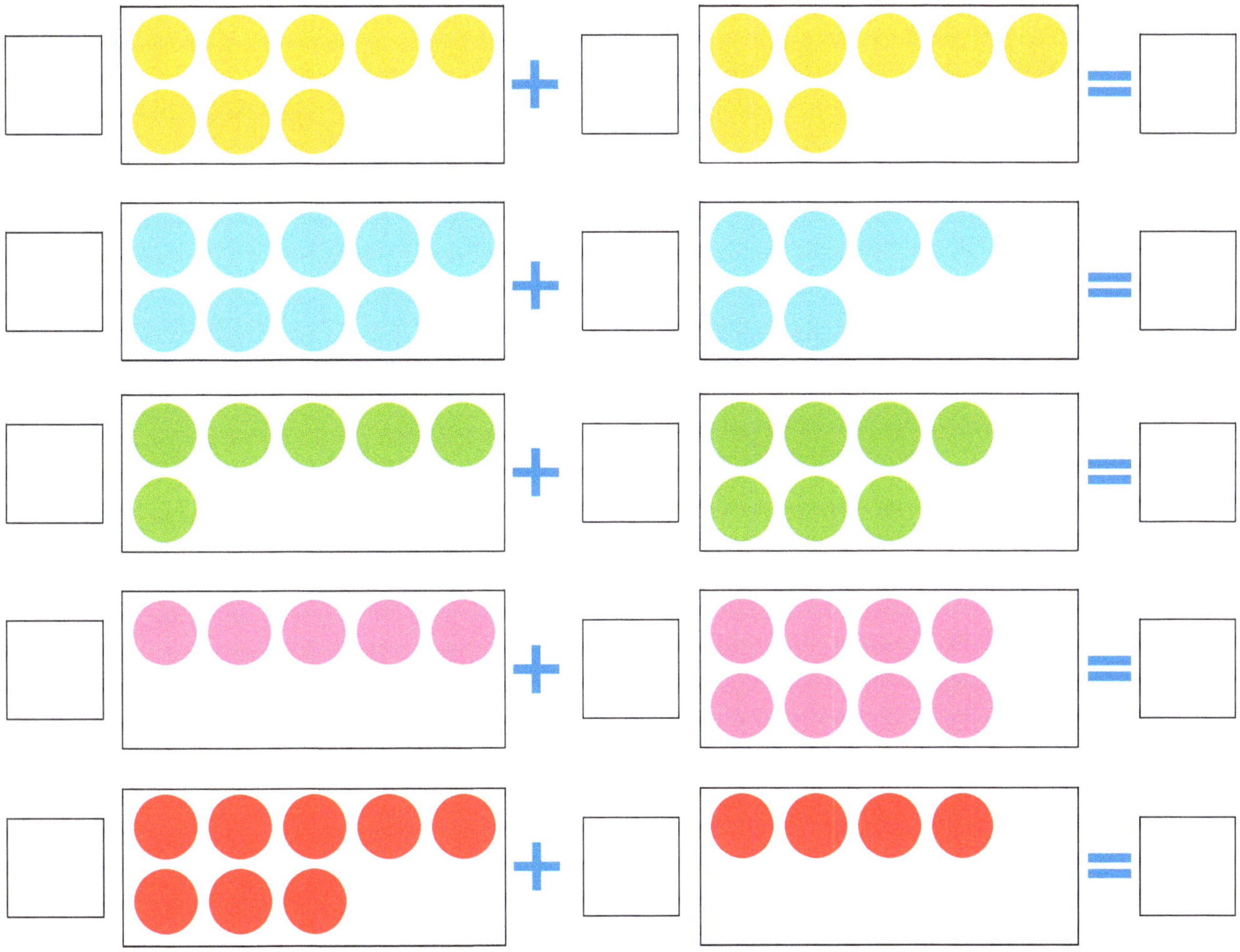

LESSON 60 Practice Adding with Blocks

God, we praise You for giving us a heart that seeks to learn.

Memory Work — *Practice weighing items on a scale.*

Add these blocks together to get one bigger chunk of blocks. Add the numbers and write the answer in the box.

___ + ___ = ___

___ + ___ = ___

___ + ___ = ___

☐ + ☐ = ☐

☐ + ☐ = ☐

☐ + ☐ = ☐

Choose your own blocks and add them together yourself. Start with the blocks in the picture and add more to them. Complete your own addition exercise.

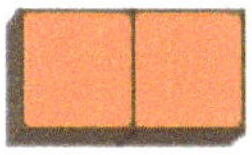

☐ + ☐ = ☐

Addition Activity

Practice using your math blocks along with a set of UNO™ cards. Flip over two cards with numbers. Find the math blocks with the same numbers. Add the two numbers together.

LESSON 61

More Practice Adding

God, we praise You for showing Your faithfulness in addition.

Memory Work — *Practice counting to 30. Can you count backward from 30 too?*

Add each row and write the number in the box. Use a number line or blocks to add these together.

$6 + 4 = \square$ $3 + 9 = \square$

$9 + 2 = \square$ $4 + 5 = \square$

$5 + 8 = \square$ $7 + 2 = \square$

$4 + 0 = \square$ $1 + 8 = \square$

$9 + 6 = \square$ $5 + 3 = \square$

$7 + 3 = \square$ $6 + 9 = \square$

Continue to add, matching the question balloon to its colored answer balloon. You may jump on a number line to help you with the adding.

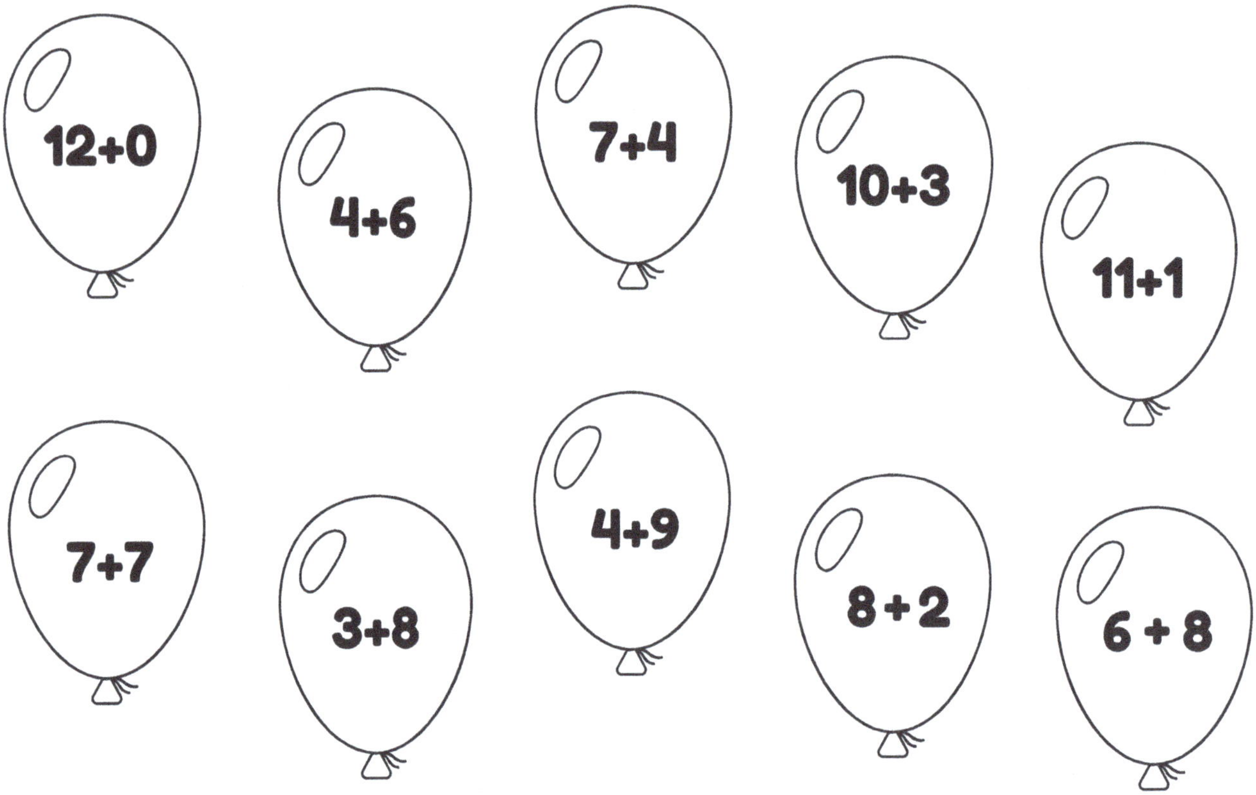

Let's Play!

Play with dominos. Connect several dominos end to end. Write addition equations for the numbers on each domino. Solve your equations.

Get Crafty!

Access addition activities using craft #1 at **Generations.org/kmath**.

LESSON 62

Even More Practice Adding

God, help us to work hard at math so that we may glorify You!

Memory Work — *Practice telling time.*

The children are flying through addition now that they have been practicing each day. Where do you think they are going? Where would you like to visit someday?

$7 + 2 = \square$	$8 + 1 = \square$
$4 + 4 = \square$	$2 + 2 = \square$
$6 + 1 = \square$	$4 + 3 = \square$
$0 + 8 = \square$	$5 + 3 = \square$
$2 + 5 = \square$	$9 + 0 = \square$

Sometimes math is written sideways and sometimes it is written up and down. Make sure you always look for the symbol that tells you what to do. The plus (+) symbol shows us that we need to add.

$$\begin{array}{r} 0 \\ +1 \\ \hline \end{array} \qquad \begin{array}{r} 0 \\ +1 \\ \hline \end{array} \qquad \begin{array}{r} 2 \\ +1 \\ \hline \end{array}$$

$$\begin{array}{r} 0 \\ +2 \\ \hline \end{array} \qquad \begin{array}{r} 2 \\ +1 \\ \hline \end{array} \qquad \begin{array}{r} 2 \\ +2 \\ \hline \end{array} \qquad \begin{array}{r} 3 \\ +4 \\ \hline \end{array}$$

$$\begin{array}{r} 0 \\ +3 \\ \hline \end{array} \qquad \begin{array}{r} 1 \\ +3 \\ \hline \end{array} \qquad \begin{array}{r} 2 \\ +3 \\ \hline \end{array} \qquad \begin{array}{r} 3 \\ +5 \\ \hline \end{array}$$

$$\begin{array}{r} 0 \\ +4 \\ \hline \end{array} \qquad \begin{array}{r} 4 \\ +1 \\ \hline \end{array} \qquad \begin{array}{r} 2 \\ +4 \\ \hline \end{array} \qquad \begin{array}{r} 4 \\ +4 \\ \hline \end{array}$$

Play It!

Access a bingo addition game (game #9) at **Generations.org/kmath**.

LESSON 63

Making Smaller Numbers

God, thank You for taking away our sin so that we can come close to You!

Memory Work — *Practice counting backward from 30.*

And you know that [Jesus] was manifested to take away our sins... (1 John 3:5)

When we add things together, we get bigger numbers. But now we will learn to subtract. That means we will take away from a bigger number to get a smaller number.

There were 3 little birds in this nest. Then, 1 bird flew away. Subtract 1 bird from the 3 birds that were in the nest. How many birds are left in the nest? Remember, the adding sign looks like this: +. The subtracting sign looks like this: – (or we say, "minus"). We say, "Three minus one equals two." Can you say that?

$$3 - 1 = 2$$

Let's start with 2 bananas. You eat 1 banana. How many are left?

Now, imagine that you have 4 lollipops, and you eat 1 of them. How many do you have left?

$$4 - 1 = 3$$

Four minus one equals three. There are 3 left that you can share with your family.

Let's subtract!

Count the bananas. How many bananas did you start with? How many bananas did you take away? How many bananas are left? Write the number for each group of bananas in the boxes.

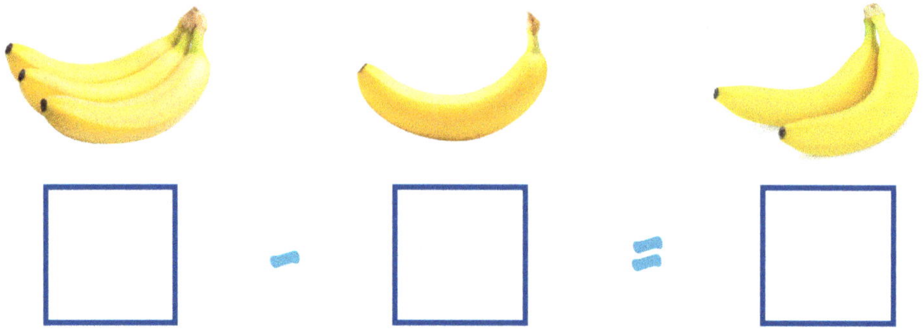

Solve these subtraction problems by making an X on the number of items you are subtracting. Count how many are left over and write the answer.

Take away 3 apples. Eat them! How many are left?

$5 - 3 =$ ☐

Take away 2 balloons. Pop them! How many are left?

$5 - 2 =$ ☐

Take away 1 milkshake. Drink it! How many are left?

$3 - 1 =$ ☐

Take away 3 colored pencils! Draw lots of pictures with them! How many pencils are left?

$6 - 3 =$ ☐

LESSON 64

Practice Subtracting

God, help us to subtract those things that take our eyes off You.

Memory Work — *Practice counting money.*

Subtract the socks. Remember to count what is left over.

4 - 1 =

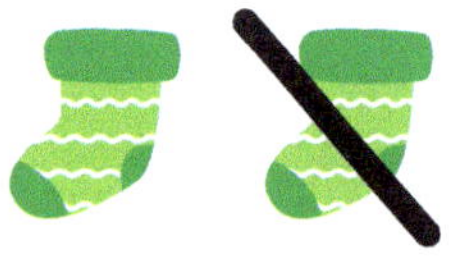

2 - 1 =

4 - 2 =

3 - 2 =

4 - 3 =

How many of these birds flew away?

Count the birds in the cage. Then, subtract the birds that flew away. How many are left? Write the answer in the box. (Hint: you can cross off the birds in the cage that flew away to help you count.)

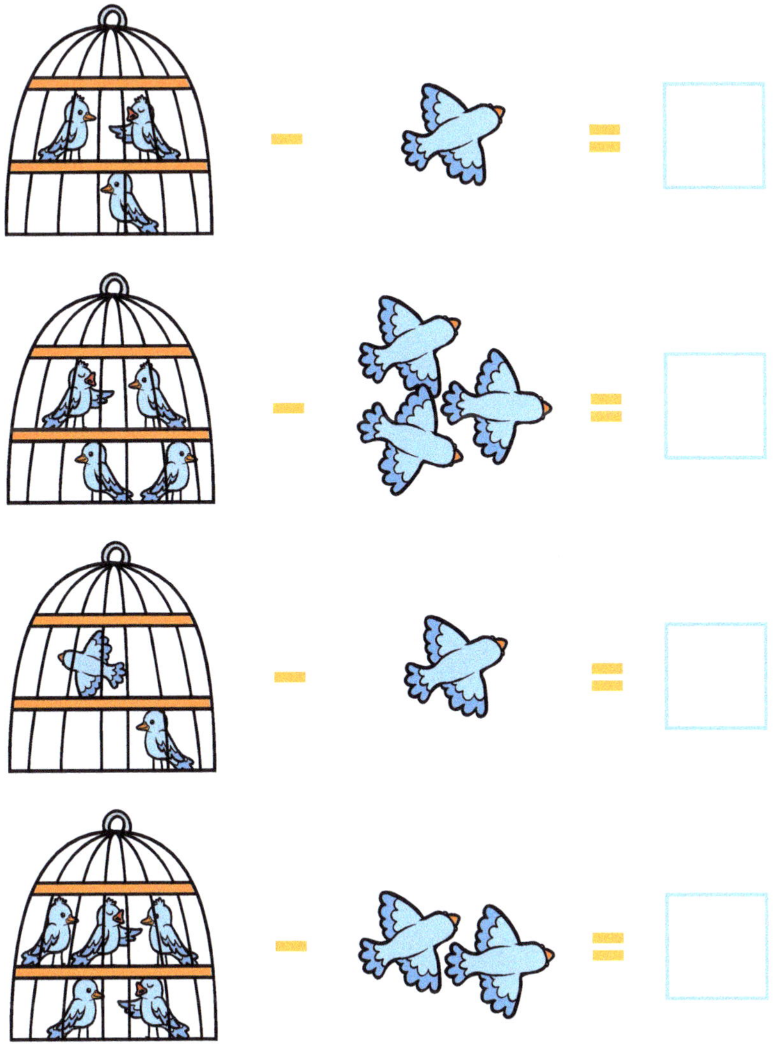

Watch It!

Learn subtraction with dinosaurs. Access video #22 at **Generations.org/kmath**.

Subtract. Color the piece of the bike with the color that matches the answer.

LESSON 65

More Practice Subtracting

God, help us to remove selfishness so that we can be more like You.

Memory Work — *Practice counting from 1 to 25 forward and backward.*

Subtract the sweet treats that you want to share with friends. Write the answer in the box.

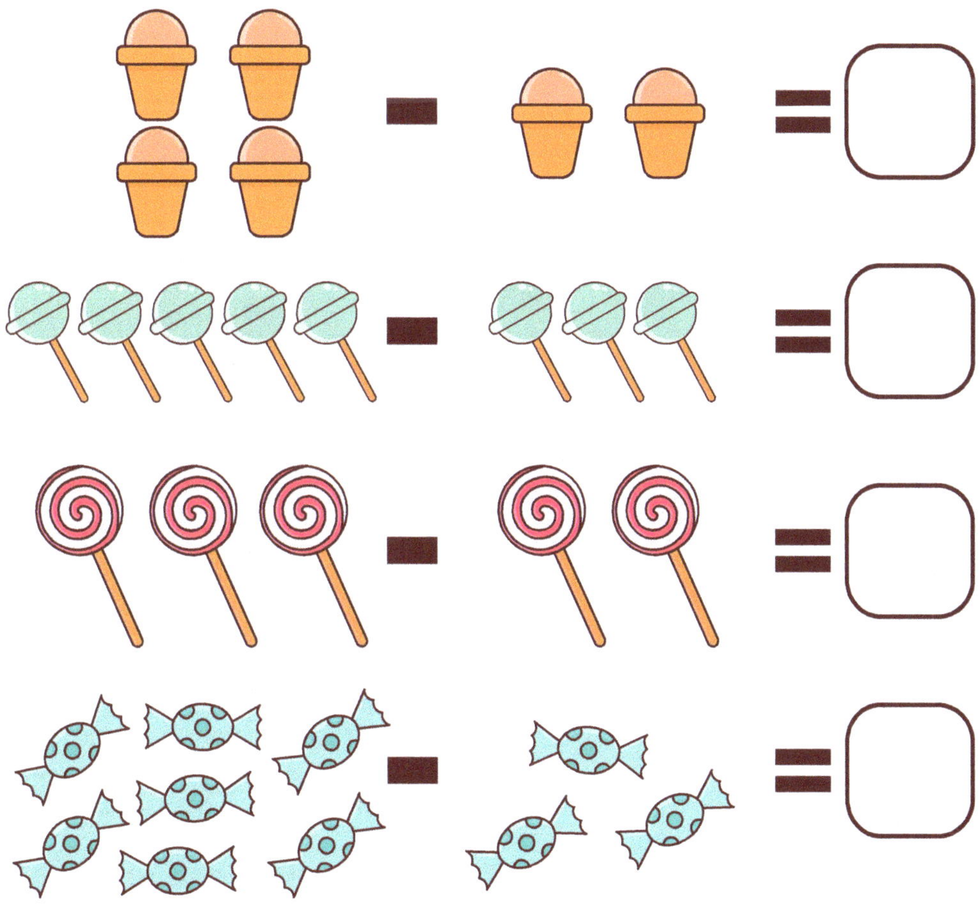

Subtract the popsicles.

$4 - 2 = \square$

$3 - 2 = \square$

$5 - 2 = \square$

$5 - 4 = \square$

$2 - 1 = \square$

Time to Play!

Have a parent write out subtraction problems on index cards. Roll up a small ball of playdough for each number you start with. Smash the playdough balls that you are subtracting. Count the rest to see the answers.

LESSON 66 Addition and Subtraction!

God, we worship You because You always know the answers!

Memory Work — *Practice measuring liquids.*

Solve and Color.

Pay attention! There are both addition and subtraction problems!

Solve the equation for each key. Notice that some have a plus sign (+) and some have a minus sign (–). Draw a line to connect the answer for the key to the number on the apartment building.

Run and Sort!

Label some brown paper bags or gift bags 2 through 5. Set them in a line a short distance from each other. Have your parent write subtraction and addition questions on individual index cards. The answers will be 2 through 5, but they won't be written on the cards! Race yourself to find the answer and place the card in the right bag. Have your parent choose a starting line and time how long it takes you to go through the pile of cards. Race against your own time or against a family member.

Subtraction Bowling!

Play a subtraction bowling game! View game #10 at **Generations.org/kmath**.

CHAPTER 8

God Made Patterns

LESSON 67 Patterns in Creation

God, thank You for shaping things in the way they need to go!

Memory Work — *Practice drawing triangles and squares. Try to draw shapes with more sides.*

> "And see to it that you make them according to the pattern which was shown you on the mountain." (Exodus 25:40)

God made patterns! Patterns are everywhere in nature. Do you see many of the same things in these pictures? A pattern repeats the same beautiful design over and over again. The patterns may involve different shapes or different colors. They may appear on plants or animals.

What is the pattern of colors on the zebra below? That's right! It is black and white and black and white, over and over again.

What pattern of colors did God create for the tiger?

What pattern do you see on this monarch caterpillar?

God also made the pattern of seasons for us. Can you name each of these?

See It in Nature!

Take a quick nature walk and see how many patterns you can find in things that God created! Make it a game by seeing who in your family can spot and explain the pattern first.

Spot It!

Butterflies usually have the same pattern on each of their wings. Sometimes that pattern may be repeated for the same type of butterfly. Can you find two identical butterflies in the picture? Color the matching butterflies using the same colors. Color the remaining butterflies differently.

What comes next in the weather pattern? Draw the correct picture in the empty box.

Here are some bugs you found walking in a row. Notice the pattern that they make and draw the next bug in the empty box.

LESSON 68 Patterns of Colors

God, thank You for the colorful world of patterns You made for us.

Memory Work — *Practice naming colors including light and dark versions of the colors.*

Watch It!

Learn about patterns with videos #23 and #24 at **Generations.org/kmath**.

What color square should come next? Draw and color the picture in the empty box.

Look carefully at the pattern of the triangles. When the pattern continues, what color triangle comes next? Draw and color the triangle in the empty box.

Notice the different shapes with different colors. Continue the pattern and try to draw the correct shape in the empty box. Color it with the next color in the pattern.

LESSON 69 Patterns of Numbers

God, we praise You for the order of numbers. Help us to use the tools of math wisely.

Memory Work — *Practice telling time.*

Teammates usually wear different numbers on their uniforms. Check the players 1 through 10 to see which player is missing. Circle the missing number on the shirt at the bottom of the page.

Fill in the Missing Numbers

Look at the pattern of the numbers counting 1 through 10 from top to bottom. Fill in the missing numbers on the apples.

Fill in the missing numbers here. What comes between 2 and 4? What comes between 5 and 7? What comes after 9?

1 2 ___ 4 5 ___ 7 ___ 9 ___

Find the pattern and fill in the missing numbers for the rest of these exercises.

6 7 8 6 7 ___ 6 ___

4 5 4 ___ 4 5 ___ ___

3 2 ___ 3 ___ 1 3 2 ___

Fill in the missing numbers on the car windows. What number comes before the number on the middle window? What number comes after it? The first green car is already filled in for you. You can look back at a number line to help you find the missing numbers for each of the cars.

Game Time!

Try this counting game. Dino sorts lets you count as you place the pieces in order onto the dinosaur. Start off by playing a few rounds counting by 1. If you want a challenge, try playing again counting by 2 or 5. You may want to look at a number line if you are playing the challenging levels. Access game #11 at **Generations.org/kmath**.

LESSON 70 Lots of Patterns

God, we praise You for the variety of things to learn in this world!

Memory Work — *Practice counting to 30, forward and backward.*

The number line can help us find patterns. Look at how many jumps it takes to go from one dot to the next. Find the next 3 numbers to continue the pattern.

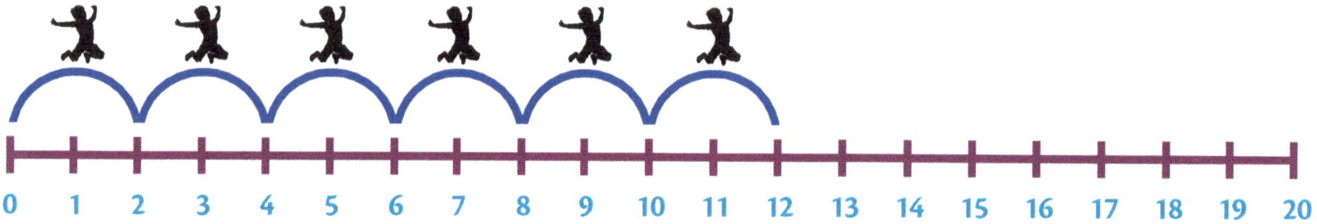

0, 2, 4, 6, ____, ____, ____

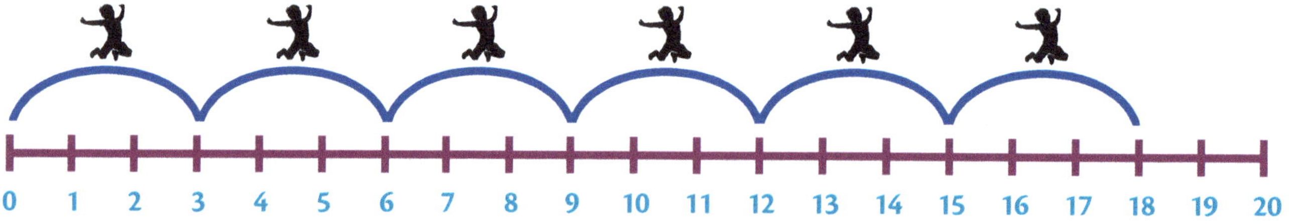

0, 3, 6, 9, ____, ____, ____

Add jumps of your own to show the last three jumps here!

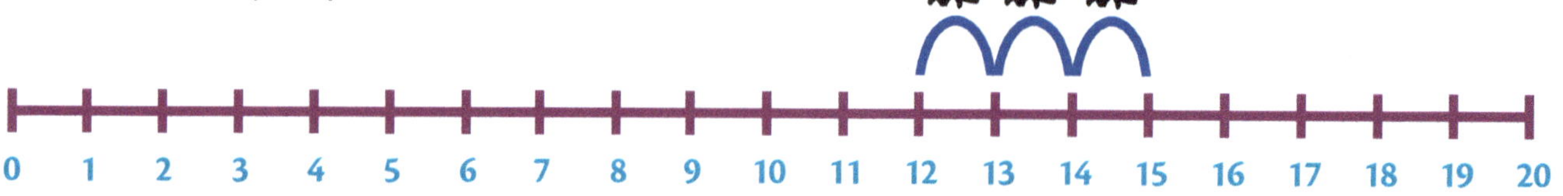

12, 13, 14, 15, ____, ____, ____

What comes next on the outdoor pattern? Draw the correct picture in the box.

Complete the animal pattern by telling your parent which animal comes next.

Drawing from a Pattern!

Use the pattern to copy the picture. Hint: Look at 1 square at a time and try to copy what is in that square.

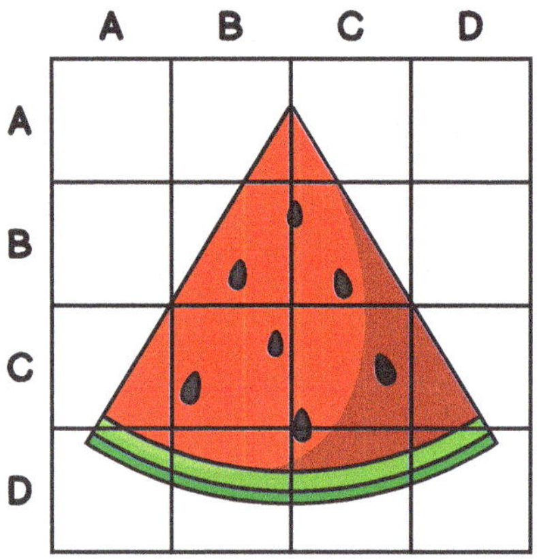

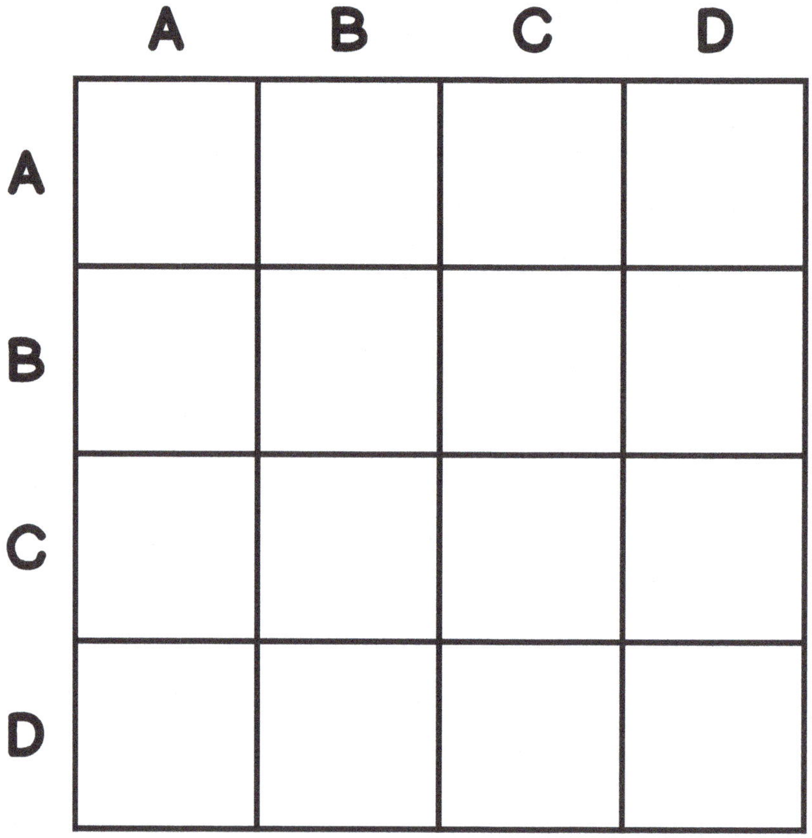

Pattern Fun with a Parent!

Try this clapping game to see if you can master the pattern with clapping. Can you create your own pattern? Access game #12 at **Generations.org/kmath**.

LESSON 71

Making Smaller Numbers (6 to 10)!

God, we praise You for giving us the Bible!

Memory Work — *Practice arranging 5 containers in your kitchen from shortest to tallest and lightest to heaviest. Is the order the same for height and weight?*

> "You shall not add to the word which I command you, nor take from it, that you may keep the commandments of the Lord your God which I command you." (Deuteronomy 4:2)

God's Word is perfect. It is completed for us. God doesn't want us to add anything to it. And He doesn't want us to take anything out of it. Let us be sure to read the Bible every day.

Watch It!

Practice adding and subtracting with a number line. Access video #25 at **Generations.org/kmath**.

Subtraction Time!

Let's take away some bigger numbers today! Remember, this is called "subtracting!" Practice subtraction by crossing off the number that is being taken away. Count what is left over. Write your answer in the box.

10 – 5 = ☐

8 – 6 = ☐

9 – 4 = ☐

7 – 1 = ☐

Follow the addition and subtraction signs. Color the answer with the matching color.

LESSON 72

Practice Making Smaller Numbers (6 to 10)!

God, thank You for the many things You bless us with!

Memory Work — *Practice counting money.*

We can be thankful for so many things that God has given us. Do you have a lot of teddy bears? Or books? Or pets?

If you do, remember to praise God for those blessings. Maybe you have too many of something and you'd like to share with others.

Write the number of the objects you see on the left side in the first circle. Take away some of them. Write that number in the middle circle. Then write the number of objects you see on the right side in the last circle.

○ – ○ = ○

○ – ○ = ○

Let's go shopping!

Do the adding and subtracting to find the price of each item. Write the answer on the tag beside each food item.

Parent or Sibling Challenge!

Play addition and subtraction checkers with your parent or sibling. Simply print out the checker math game board. Find some checker pieces or make your own. Do the math as you play. You may want to have a number line with you to help you solve. Access game #13 at **Generations.org/kmath**.

LESSON 73 More Practice Making Smaller Numbers (6 to 10)!

God, we continue to work hard at math to honor You!

Memory Work — *Practice measuring weight using a scale.*

Learning subtraction takes work. We practice and practice so that it will become easier and easier. Soon you will be able to handle harder and harder math and you will be glad you practiced!

Help Buster the cat subtract these numbers. Write the answer in the box.

5 − 1 = ☐	6 − 4 = ☐
4 − 2 = ☐	7 − 6 = ☐
3 − 0 = ☐	9 − 2 = ☐
9 − 5 = ☐	4 − 1 = ☐
8 − 8 = ☐	8 − 5 = ☐

Count and Subtract!

God made dogs, cats, and people too. Write the number of these wonderful creations you see on the left side in the first circle. Take away some of them. Write that number in the middle circle. Then write the number of dogs, cats, or people you see on the right side in the last circle.

○ − ○ = ○

○ − ○ = ○

○ − ○ = ○

Complete the mathematical house. Talk to your parent about what rooms should go in each colored section.

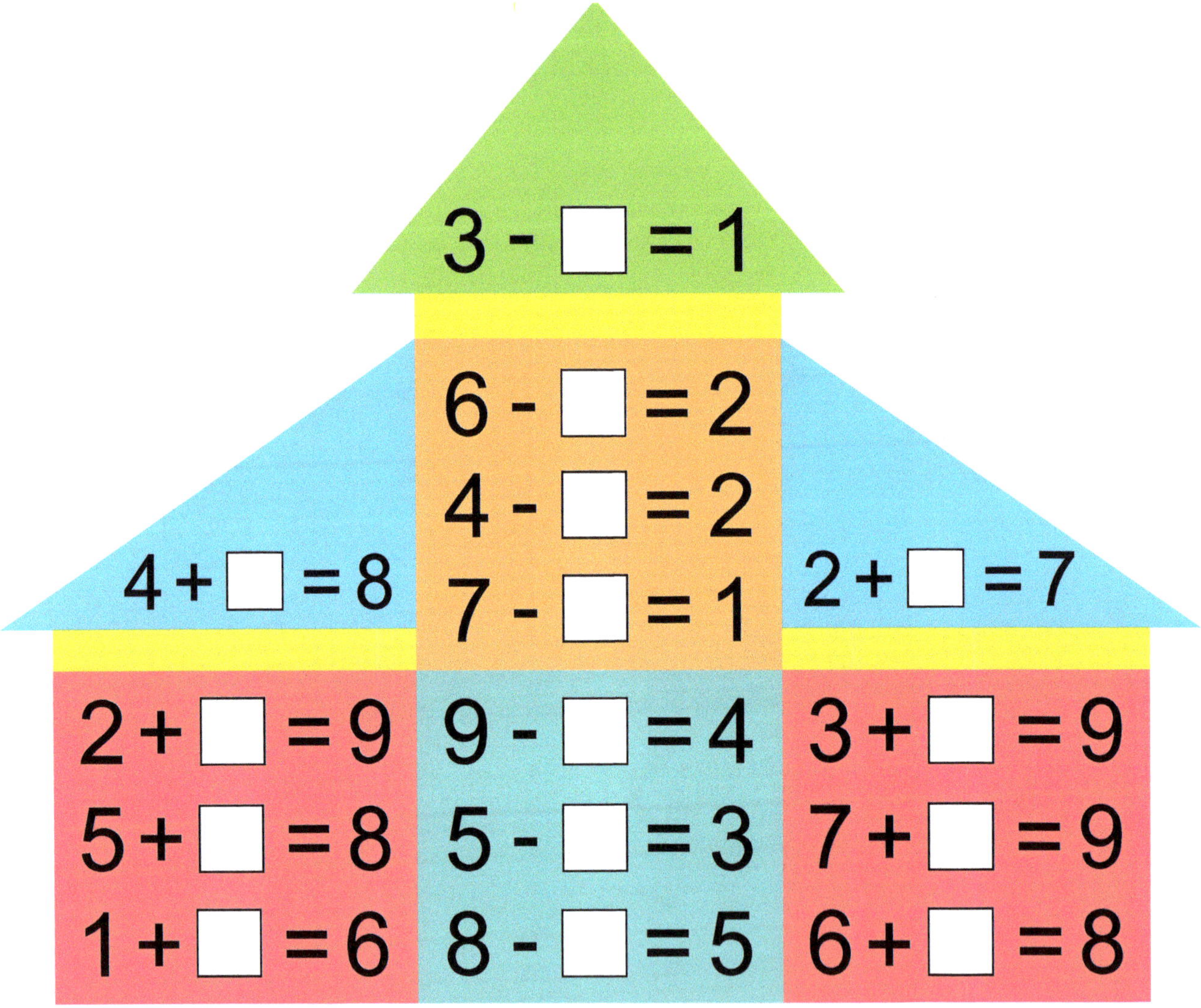

Let's Get Crafty!

Try the subtraction flower activity to make a beautiful display of what you are learning. You will need a marker, some colored paper, scissors, and glue. Ask your parent where you can hang your finished flowerpots. Access craft #2 at **Generations.org/kmath**.

LESSON 74 Challenges Making Smaller Numbers (6 to 10)!

God, math can be challenging at times, but we welcome the chance to learn!

Memory Work — *Create a few series of patterns with colors and shapes.*

Today, we will try a few fun puzzles. These puzzles will make you think.

Which number will you match with the penguin, with the hedgehog, and then with the squirrel? Match a number which each animal.

Solve It!

Now let's try a math crossword puzzle. Fill in the missing numbers as you add or subtract. Pay attention to the + and – signs.

Basketball Subtraction!

Play basketball as you do math. Grab a small ball and either a laundry basket or a small, empty trash can, and set your basket somewhere high. Have your parent quiz you with flash cards, and if you answer correctly, you get to shoot the ball for an extra point. Make it a competition with your siblings.

Maze Fun

Complete the maze by solving the problems. Connect the answers that are in counting order. Example: 1, 2, 3...

CHAPTER 9

Counting Faster

LESSON 75

Grouping: Fast Counting by 10s

God, thank You for giving us ways to organize things in this world.

Memory Work — *Practice counting to 30, backward and forward.*

> And Moses chose able men out of all Israel, and made them heads over the people: rulers of thousands, rulers of hundreds, rulers of fifties, and rulers of tens. (Exodus 18:25)

God created so much for us to enjoy. There is so much to see and to count. We can count by groups to count many things at the same time.

Today, we're going to learn to count big numbers fast!

You've already learned names for some of these big numbers. Let's practice and learn some more!

Ten, twenty, thirty, forty, fifty, sixty, seventy, eighty, ninety, one hundred.

10, 20, 30, 40, 50, 60, 70, 80, 90, 100

Notice the first digit in those numbers: 1, 2, 3, 4, 5, 6, 7, 8, 9, 10. You already know how to count from 1 to 10. When you write these numbers, write a 0 after the number.

Remember every group represents 10 items.

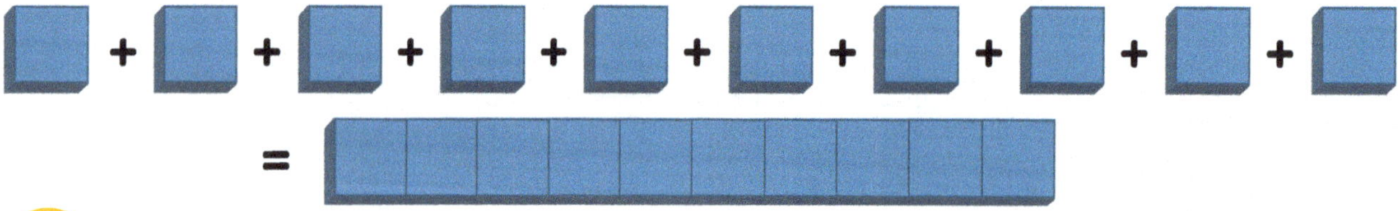

Here is how to count things really fast. We group them in tens first. Here are groups of apples. Let's count by tens! How many apples are there altogether?

10

20

30

Now look at all these condor birds! How many birds are there? Circle each group of 10 birds.

How many groups are there? ________

How many birds are there altogether? Count by 10s. ________

Let's Exercise!

Exercise while you practice counting by 10s. What else can you imagine yourself doing while you count? Access video #26 at **Generations.org/kmath**.

Sing It!

Here is more practice counting by 10s. Video #27 at **Generations.org/kmath**.

Now we can count the same way with our blocks. Count the groups of 10 blocks, and write the number in the box.

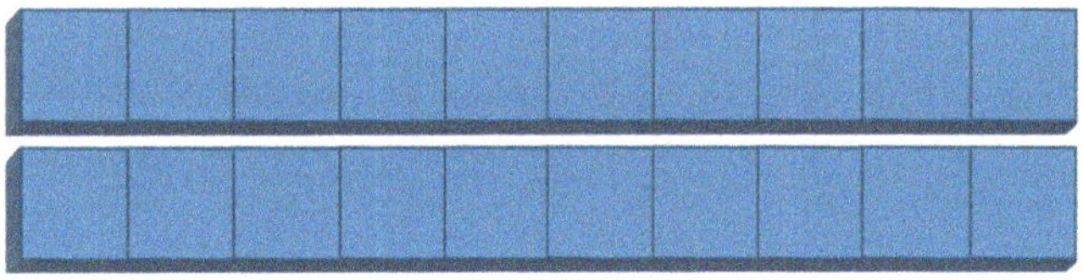

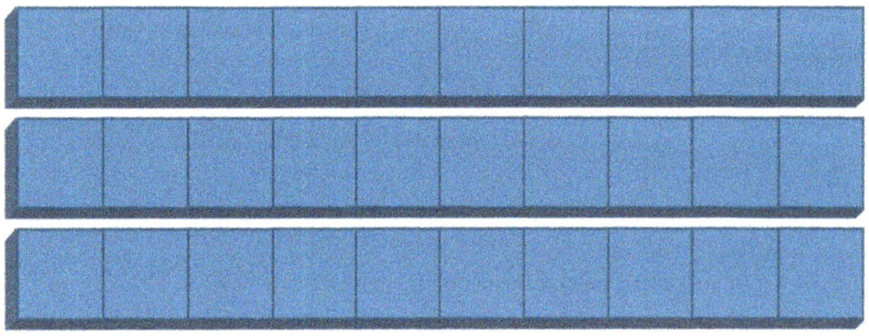

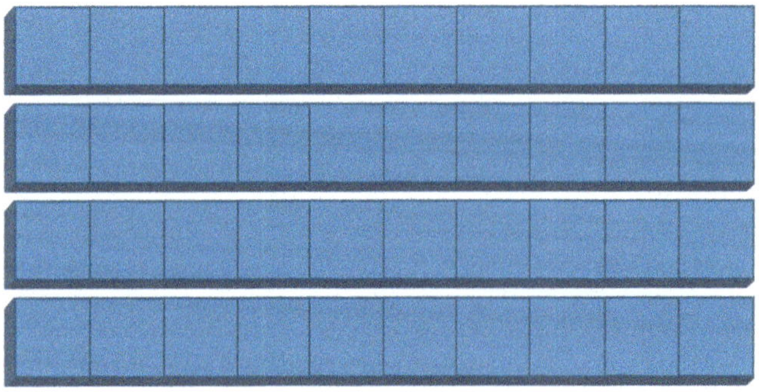

LESSON 76

Practice Counting by 10s

God, thank You for the family and friends that are able to help us learn math.

Memory Work — *Practice adding using your flash cards.*

Let's practice counting by 10s. Point to the numbers and count out loud.

Fill in the missing numbers as you count by 10s.

10	20	30	40	50		70	80	90	100
10		30	40	50	60	70	80	90	100
10	20	30	40	50	60	70	80		100
10	20	30		50	60	70	80	90	100
10	20	30	40	50		70	80	90	100
10	20	30	40	50	60	70		90	100

Fill in the missing numbers on the hearts.

Connect the dots and color!

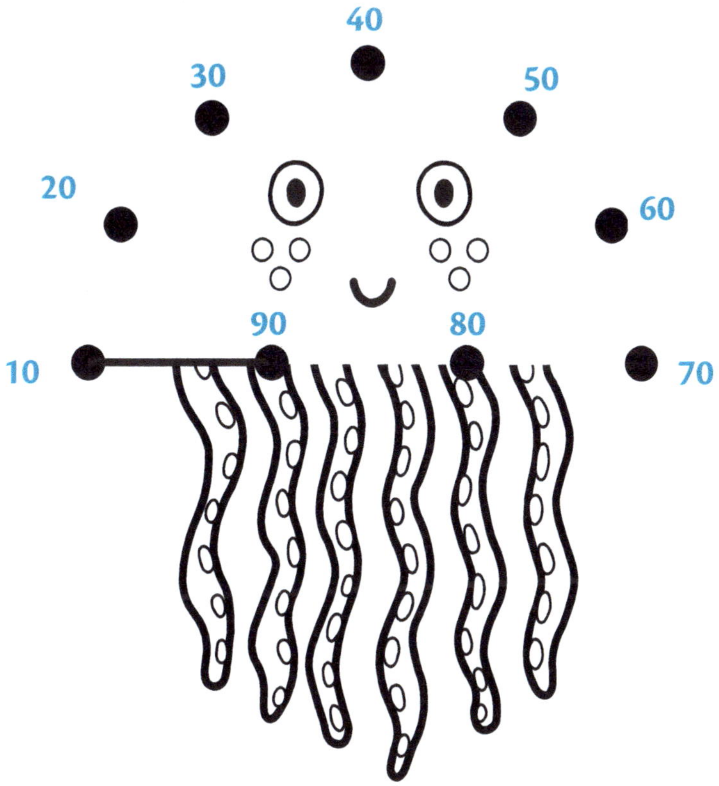

Watch It!

Count by 10s in this video. You will see groups of things you can count and then count even higher. Access video #28 at **Generations.org/kmath**.

Unscramble It!

Parent instructions: Write sets of numbers counting by 10s on a piece of paper. Cut out the numbers and fold them up. Place the papers into a balloon. Blow up the balloon with the papers inside and tie it. Have the student run, pop the balloon, and quickly place the papers in order. If more than one student is available, make it a race to see who can put them in order the fastest.

LESSON 77

Practice Adding and Subtracting

Lord, help us to serve others with our math skills.

Memory Work — *Practice creating patterns with colors and shapes.*

Try adding by memory. Adding blocks might help you too! Write the answer in the box.

1 + 1 = ☐

1 + 2 = ☐

1 + 3 = ☐

1 + 4 = ☐

2 + 1 = ☐

2 + 2 = ☐

2 + 3 = ☐

2 + 4 = ☐

Try this subtracting by memory! Taking away the blocks may help you! Write the answer in the box.

2 - 1 = ☐

3 - 1 = ☐

4 - 1 = ☐

5 - 1 = ☐

2 - 2 = ☐

3 - 2 = ☐

4 - 2 = ☐

5 - 2 = ☐

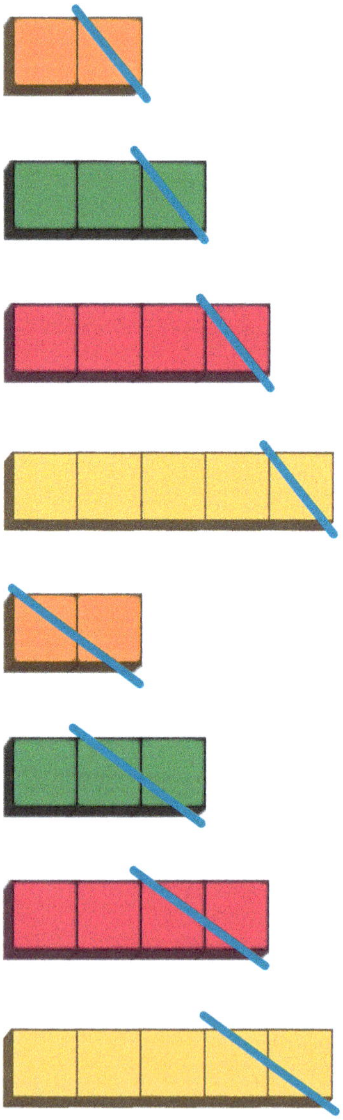

LESSON 78 Grouping: Fast Counting by 5s

God, we thank You for the small groups that help us to know You!

Memory Work — *Practice fast counting by 10s.*

The Torah, or Pentateuch, is the first 5 books of the Old Testament. Those books are Genesis, Exodus, Leviticus, Numbers, and Deuteronomy. God gave us this group of five to teach us about Creation and His laws. We can learn more about Him through these books.

Now we will learn to fast count by fives.

How many oranges do you see in this box? There are 3 rows of 5 oranges. Point to each row while you count. Now let's count the oranges, row by row… 5, 10, 15. That's right! There are 15 oranges in the box. There's a fast way to count!

Circle each group of 5 ladybugs. How many groups do you see? ________

How many ladybugs are there altogether? ________

Now, let's count 50 tomatoes using the fast count by 5s. Say the numbers out loud!

Five, ten, fifteen, twenty, twenty-five, thirty, thirty-five, forty, forty-five, fifty.

5, 10, 15, 20, 25, 30, 35, 40, 45, 50. Do you see a pattern in the second digit? 0-5-0-5-0-5-0-5. What numbers come after 50 in this pattern?

Notice It!

Take a look at a Bible with your parent. Turn to the Psalms. How many books is it divided into? Count by 5s as you look at the chapters. How high can you count?

Circle groups of 5. How many groups did you circle? ☐

Sing It!

Practice counting by 5s with this song. Access video #29 at **Generations.org/kmath**.

LESSON 79

Practice Counting and Adding

God, help us see groups of things that You created so that we can praise You for them.

Memory Work — *Practice measuring with a ruler.*

How many blocks do you see? Write the number in the box.

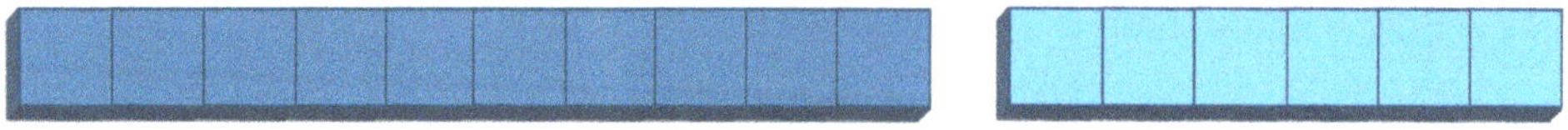

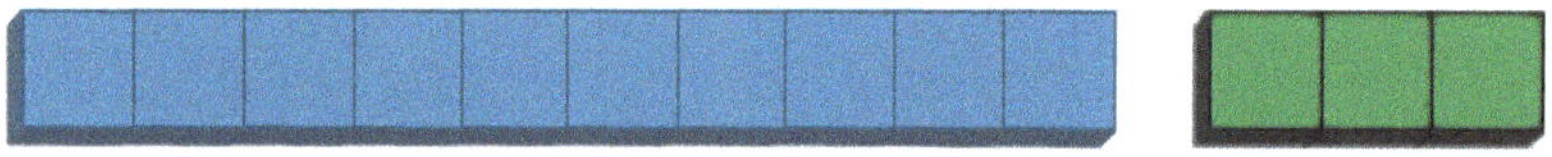

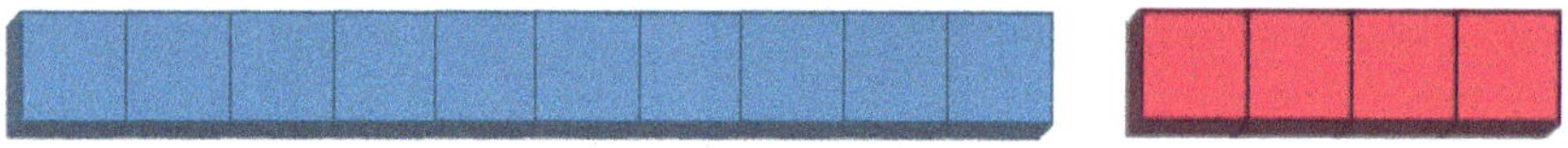

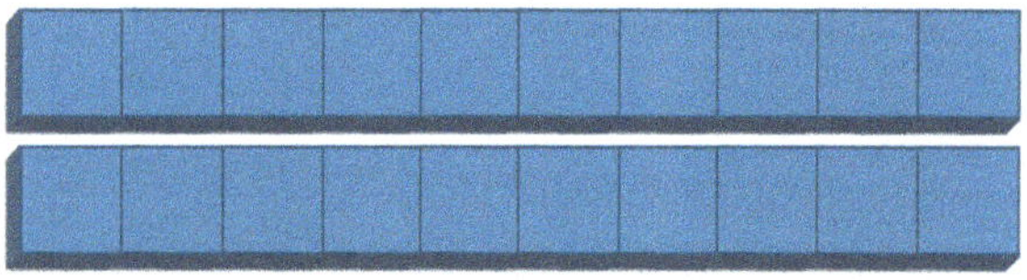

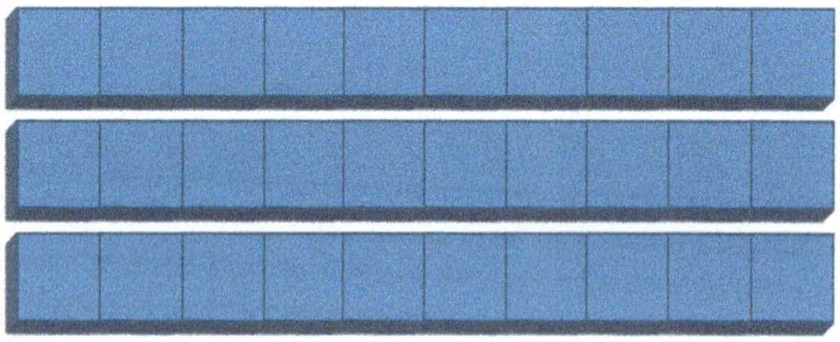

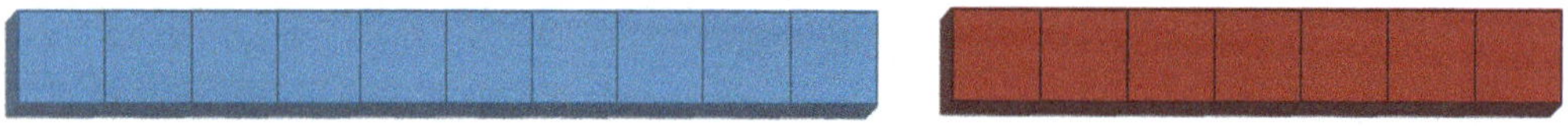

Let's add these animals together.

 +  =

 +  =

 +  =

 +  =

LESSON 80

Grouping: Practice Counting by 5s

God, nothing is missing with You. You know what we need and when we need it.

Memory Work — *Practice subtraction from 10 with your blocks.*

Count by 5s. Fill in the missing numbers.

Play It!

Here is an online game. For the Number Bubble Skip Counting, choose to count by 5s. Click on the bubble so that the number will fall into the correct treasure chest. Access game #14 at **Generations.org/kmath**.

Finish the pattern with the next number.

10 15 20 25 ☐

30 35 40 45 ☐

50 55 60 65 ☐

Make It!

Here is a number snake craft to help you remember counting by 5s: craft #3 at **Generations.org/kmath**.

Make this simple game (see game #15) with popsicle sticks. Game #1 is one you can play by yourself. Game #2 is one you can play with more people.

LESSON 81

Making Bigger Numbers

God, thank You for gathering groups together for us to count.

Memory Work — *Practice subtracting with your flash cards.*

> The Lord God, who gathers the outcasts of Israel, says, "Yet I will gather to him others besides those who are gathered to him." (Isaiah 56:8)

Have you ever been to the beach and noticed a group of birds? Another group may fly in and join those birds. Now you have a bigger group to count!

Count the objects in each group. Write the numbers down. Add the groups and write your answers. Remember, you can count all the objects together if that will help you find the answer.

7 globes + 3 globes = ☐

☐ ☐

4 lollipops + 2 lollipops = ☐

☐ ☐

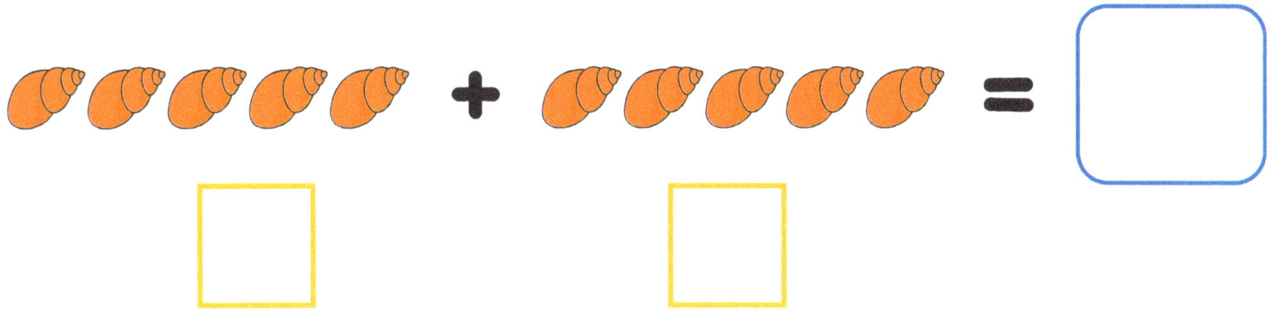

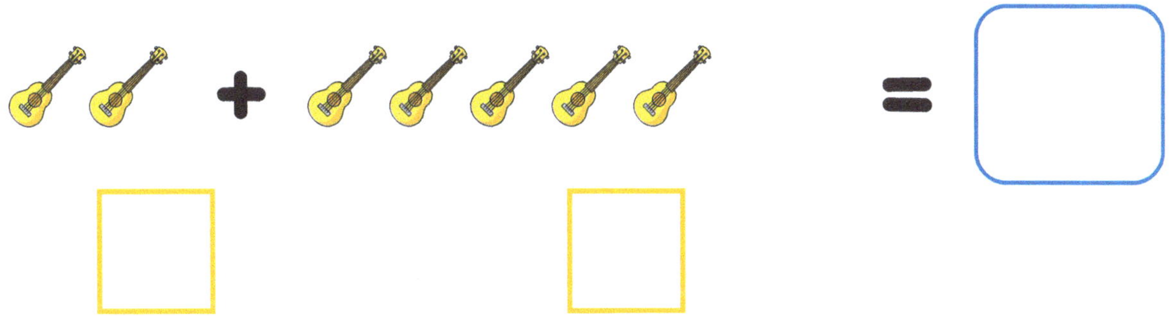

Watch It!

Count the seashells to solve the addition questions. Access video #30 at **Generations.org/kmath**.

Adding

Add 1 more to these big numbers. Use the number line if that helps you. Start with the first number, and then take a jump of only 1. Where do you end up?

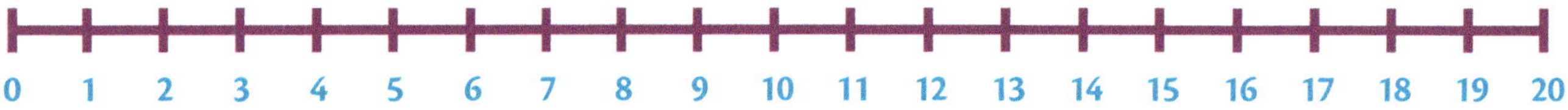

$8 + 1 = \square$

$11 + 1 = \square$

$9 + 1 = \square$

$12 + 1 = \square$

$10 + 1 = \square$

$20 + 1 = \square$

Add 0 more to these big numbers. Adding zero is no jump at all! You stay where you are.

$8 + 0 = \square$

$20 + 0 = \square$

$15 + 0 = \square$

$30 + 0 = \square$

$10 + 0 = \square$

$40 + 0 = \square$

LESSON 82

Practice Making Bigger Numbers

God, thank You for making things grow.

Memory Work — *Practice counting money.*

God gave us so many groups of delicious food to eat. Add the groups and write your answers. See if you have any real fruit at home that you can count.

Add or subtract.

Color the answer with the color shown.

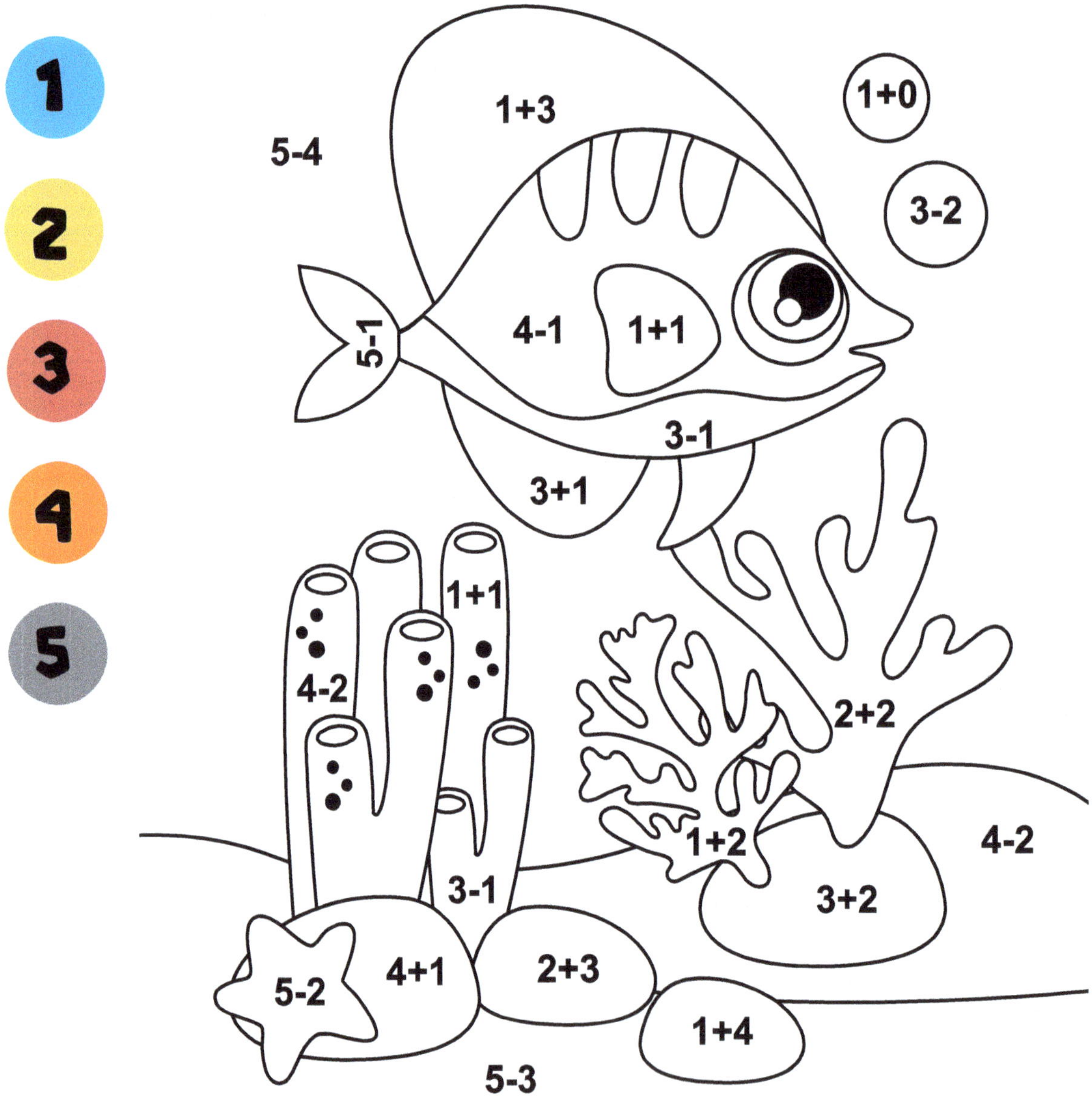

Activity Time!

Try the fishing for numbers game. Make numbers on fish. Use magnets to pick up the fish with your fishing pole. This game can be played with or without a water table. See the website for ideas on how to make your fish and gear. Access game #16 at **Generations.org/kmath**.

LESSON 83

More Practice Making Bigger Numbers

God, help us to trust in You to help us when we struggle.

Memory Work — *Practice counting by 5s and 10s.*

Count the total number of cookies and match to the answer on the right.

- 16
- 13
- 17
- 15
- 20

Compare the watermelon pieces and circle the portion that is bigger.

Craft Time!

Create watermelon mini books that show the number of seeds on the front and the math questions on the inside. See the examples for ideas. Access craft #4 at **Generations.org/kmath**.

Twister Math

Let's play Twister Math. If you have a Twister game, use sticky notes to randomly cover the circles with numbers 1 to 20. Have your parent tell you addition questions instead of colors, and you must place your hand or foot on the circle with the correct number.

If you don't have the game, try using paper plates in a 4 x 5 arrangement for the circles, and have your parent tell you which hand or foot to place on the plate. See how many questions you can answer correctly while keeping your balance!

CHAPTER 10

God's Subtraction and Sharing

LESSON 84 Subtraction on a Number Line

Father God in heaven, we thank You for adding and taking away things from us. You are always wise and always good to us.

Memory Work — *Practice counting to 30, forward and backward.*

> And Job said: "The Lord gave, and the Lord has taken away; Blessed be the name of the Lord." (Job 1:21)

The Lord gave Job 7 sons and 3 daughters, and then He took them away from Job. Then, He gave Job 7 more sons and 3 more daughters. God is always good even when He takes things away from us. Let us trust Him for this.

Today, we will take away some bigger numbers. We will start with a big number, and then take away big and small numbers. We will jump the other way on the number line. Let's start with two small jumps.

$$12 - 1 = 11$$
$$20 - 1 = 19$$

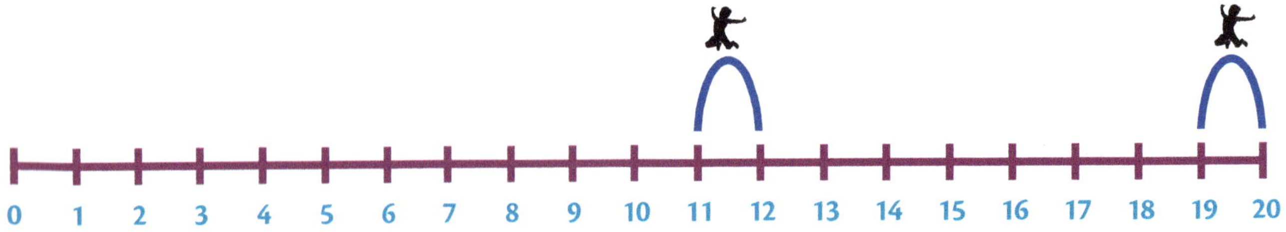

You can see the jumpers are facing the other direction. The first jumper starts at 12 and jumps 1 space down to 11. The other jumper starts at 20 and jumps to a smaller number too. He jumps from 20 down 1 space to 19.

Now try a few yourself.

$19 - 1 = \square$

$18 - 1 = \square$

$17 - 2 = \square$

$16 - 2 = \square$

$11 - 3 = \square$

$10 - 3 = \square$

When you add, jump up the hill on the number line. When you subtract take a jump down the hill! The first exercise is completed for you. Draw in each jump on the number line.

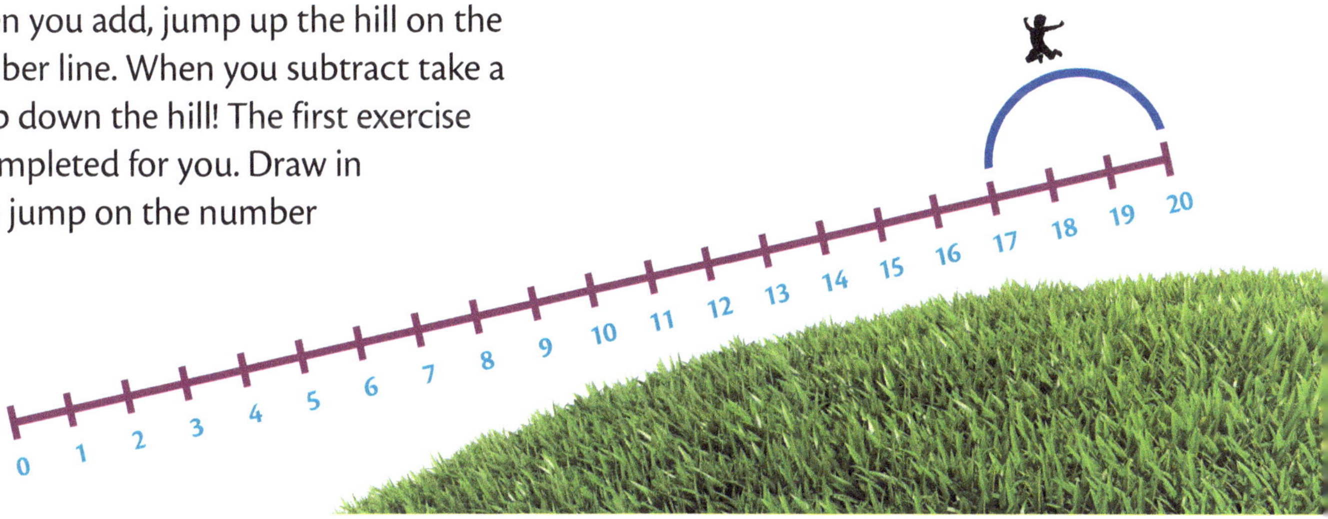

$20 - 3 = 17$

$17 - 3 = \square$

$16 + 4 = \square$

$16 - 4 = \square$

$13 + 3 = \square$

$12 - 5 = \square$

$7 + 6 = \square$

$5 - 5 = \square$

LESSON 85

Something New and Some Practice

Father God in heaven, we thank You for adding and taking away things from us. You are always wise and always good to us.

Memory Work — *Practice counting by 5s and 10s.*

Count left and right.

Look at the chicks. Some are pointing to the left (blue arrow), and some are pointing to the right (red arrow). Write the numbers of chicks pointing left and right in the box.

Sing It!

Watch the Left and Right Song (video #31) at **Generations.org/kmath**.

How long are these things?

cm

cm

cm

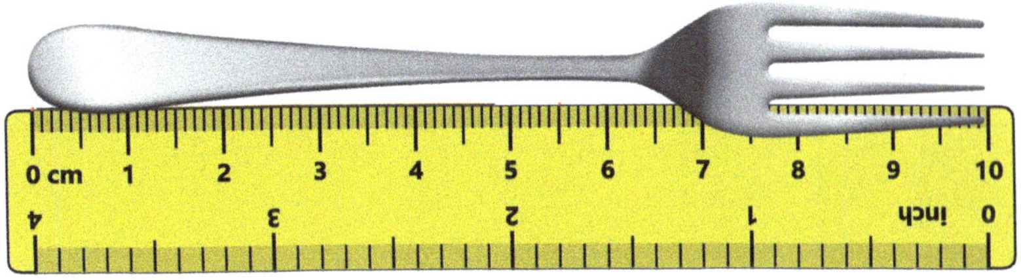

cm

Measuring Distance

Go outside and practice doing long-distance jumps. Use chalk to create a starting line that you jump from. Mark the ground where you jump to. Measure how far you jumped with a ruler. Repeat a few times. What is the longest distance you jumped?

LESSON 86

Count the Bigger Numbers to 100!

Thank You, God, for making big numbers. Help me to count higher and higher today! Amen.

Memory Work — *Practice using your flash cards to add and subtract.*

[Jesus said], "What do you think? If a man has a hundred sheep, and one of them goes astray, does he not leave the ninety-nine and go to the mountains to seek the one that is straying?" (Matthew 18:12)

That's a lot of sheep! Let's count them by 10. Where is the missing sheep?

10

20

30

40

50

60

70

80

90

100

You have practiced counting to 30, and you should have this memorized by now. Let's complete these numbers to 100.

Count 31 to 40.	31	32	33	34	35	36	37	38	39	40
Count 41 to 50.	41	42	43	44	45	46	47	48	49	50
Count 51 to 60.	51	52	53	54	55	56	57	58	59	60
Count 61 to 70.	61	62	63	64	65	66	67	68	69	70
Count 71 to 80.	71	72	73	74	75	76	77	78	79	80
Count 81 to 90.	81	82	83	84	85	86	87	88	89	90
Count 91 to 100!	91	92	93	94	95	96	97	98	99	100

Can you find **40**? Write 40 next to it.

Can you find **60**? Write 60 next to it.

Can you find **80**? Write 80 next to it.

Can you find **30**? Write 30 next to it.

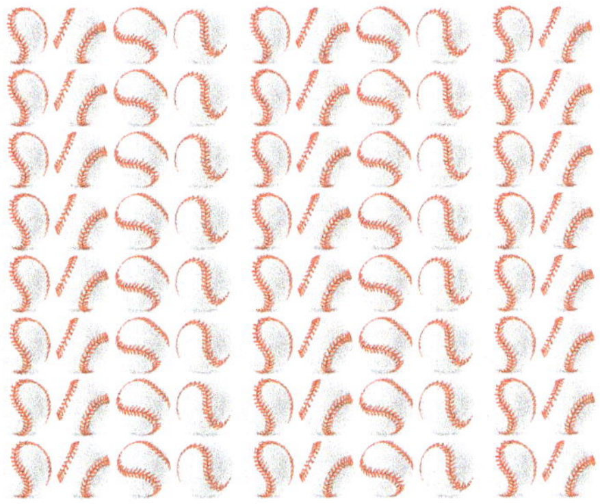

Fill in the missing numbers.

1	2	3	4	___	6	7	8	9	10
11	12	13	14	15	16	___	18	19	20
21	22	___	24	25	26	27	28	29	30
31	___	33	34	35	36	37	38	39	40
41	42	43	44	45	46	47	___	49	50
51	52	53	___	55	56	57	58	59	60
61	62	63	64	65	66	67	68	69	___
71	72	73	74	75	76	77	78	___	80
81	82	___	84	85	86	87	88	89	90
91	92	93	94	___	96	97	98	99	100

Let's Play!

Use a dice and keep adding the number you move along the path on the next page. If possible, play along with your parent and see who makes it through the path the fastest.

START
1
2
3
4
5
6
7
8
9
10
11
12
13
14
15
16
17
18
19
20
21
22
23
24
25
26
27
28
29
30
31
32
33
34
35
36
37
38
39
40
41
42
43
44
45
46
47
48
49
50
51
52
FINISH

LESSON 87

Math Fun

God, thank You that no problem is ever too big for You to handle.

Memory Work — *Practice reviewing shapes and colors.*

Let's tell a story with math!

John has 2 caterpillars. Jane has 3 caterpillars. How many caterpillars do they have altogether?

☐ + ☐ = ☐

Maria had 4 balloons. 1 blew away. How many balloons does Maria have now?

☐ - ☐ = ☐

Peter had 6 cookies. He gave 2 cookies to Leah, who didn't have any. How many cookies does Peter have left?

☐ - ☐ = ☐

4 frogs were sitting on branch. 2 hopped away! How many are left?

☐ - ☐ = ☐

Fill in the missing numbers for the patterns. (Some of these are counting up and some are counting down.)

___	1	2	3	4	5	6	7
13	14	15	16	___	18	19	20
71	72	73	___	75	76	78	79
33	34	35	36	37	___	39	40
10	9	8	7	___	5	4	3
19	18	17	16	15	14	___	12

Practice counting as you complete this dot-to-dot.

LESSON 88

More Fun Activities!

God, we praise You for bringing things out of the darkness and into the light.

Memory Work — *Practice measuring cups and pints. How many cups fill a pint?*

Draw lines to match each row of colorful gardening items with its shadow.

How many shirts? How many skirts? How many socks? How many of each kind of clothing can you find? Add the total and write the number in the box next to the item.

Add these colored eggs and put them in the same container. Color the total eggs into the last carton on the right. Be sure to write the number of each quantity of eggs in the boxes.

+

=
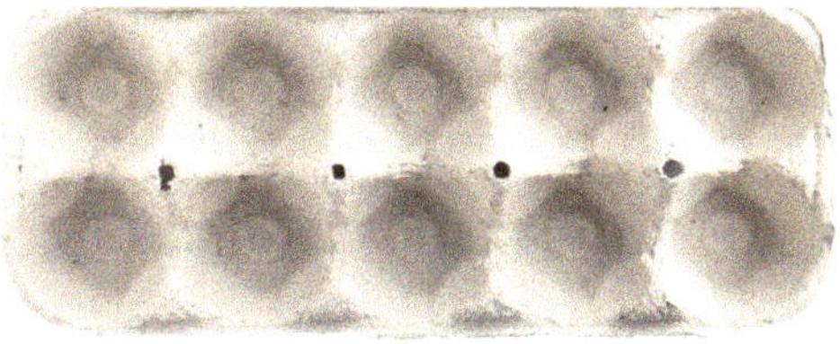

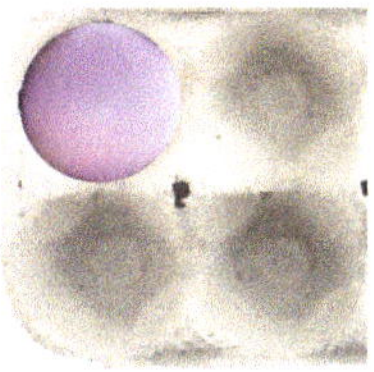
+

=
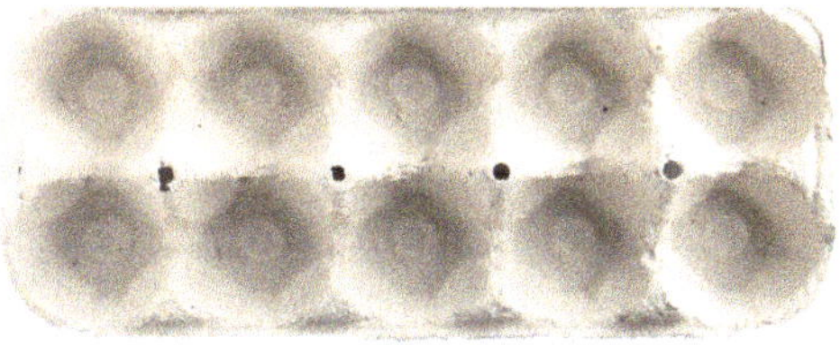

+

=
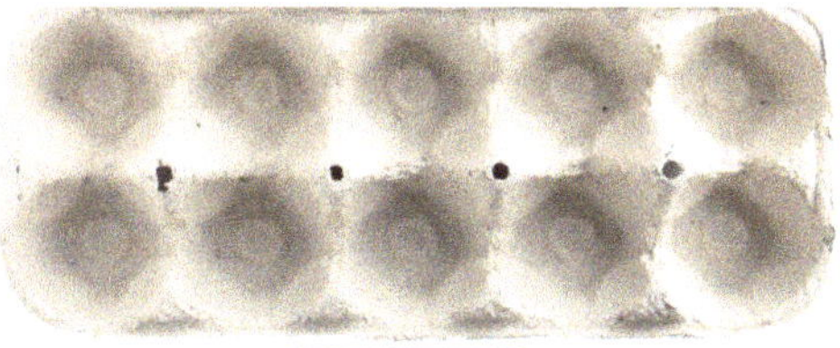

+
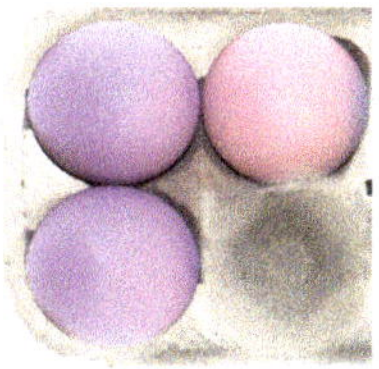
=
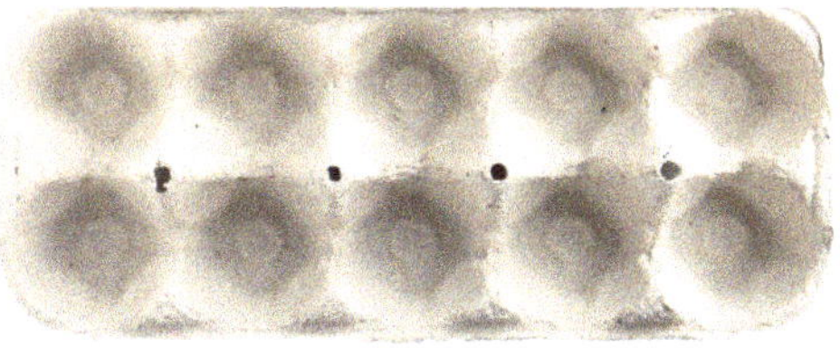

LESSON 89

Equal Parts to Share

Thank you, God, for giving us so much. Please help us to share with others.

Memory Work — *Practice counting from 1 to 30.*

> Do not neglect to do good and to share what you have, for such sacrifices are pleasing to God. (Hebrews 13:16 ESV)

God gives us good gifts. And He is so pleased with us when we share with others.

When mother gives you one big cookie, can you share it with one friend? You can cut the cookie into 2 parts the same size. We call these "equal parts." Let's cut the cookie in two pieces.

Now, suppose you cut the cookie this way so you would have the bigger piece. These are not equal parts. Maybe you should give your friend the big piece. And you could have the smaller piece! That would be a kind thing to do!

If your mother gave you one pizza, could you share it with three friends? Let's cut the pizza into 4 equal parts. Draw lines to make it four equal pieces.

Can you cut this apple into 2 equal pieces? Draw a line where you would cut it.

Share!

Share these snacks with your friends! Circle each snack that has equal parts.

Watch It!

Here is a video with examples of equal or not equal parts: watch video #32 at **Generations.org/kmath**.

LESSON 90 Practice with Equal Parts

God, thank You for giving us what is best even if it is not equal to what You give others.

Memory Work — *Practice adding big numbers on a number line.*

Circle the shapes that are cut into equal parts. Write the number of equal parts in the box near the shape you circled.

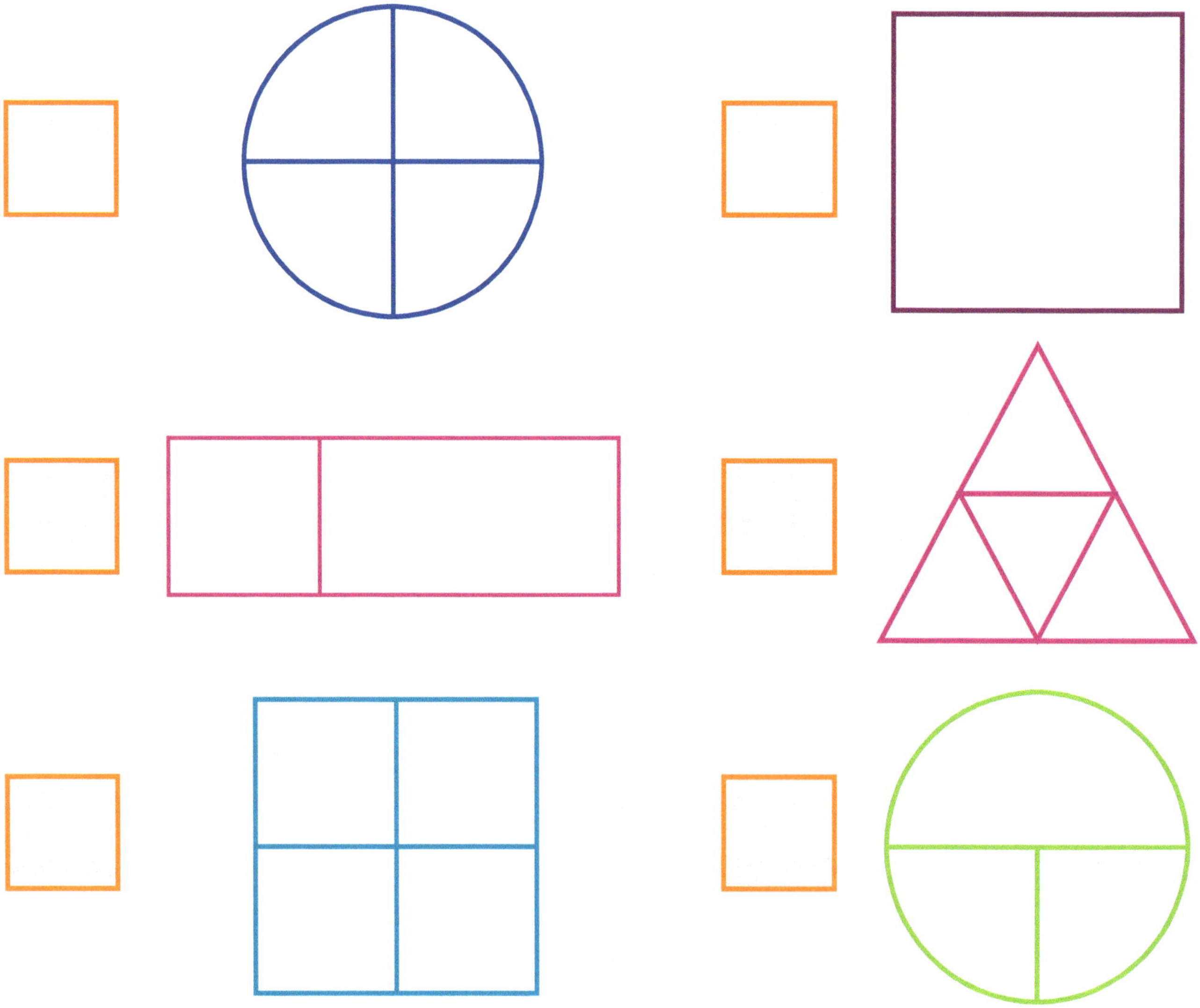

Draw 1 line to make 2 equal parts for the following shapes. Connect the dots.

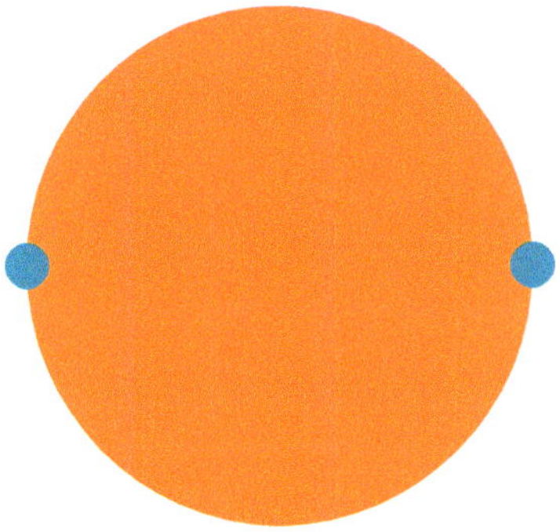

Draw 2 lines to make 4 equal parts for these shapes. Connect the dots.

Activity Time!

Practice creating equal parts using play dough or clay. Can you split it into 2 equal parts? What about 3 equal parts? How about 4 or 5 equal parts?

Help the farmer find which trees are the same. Be careful! There is only one matching pair.

LESSON 91

Final Review

God, we praise You for helping us learn math!

Memory Work — *Practice counting from 30 to 50.*

How many pieces of pie? How many equal pieces do you see here?

Let's tell a math story!

There were 8 ice cream cones. 5 children ate 5 ice cream cones. How many are left? Draw lines from each child to a cone.

$\square - \square = \square$

There were 7 bags of popcorn. 4 children ate 4 bags of popcorn. How many bags are left? Draw lines from each child to a bag of popcorn.

$\square - \square = \square$

There were 4 apples. 3 children ate 3 apples. How many apples are left? Can you figure it out without drawing lines?

☐ - ☐ = ☐

Color these blocks with two different colors (like red and blue). Then fill in the missing number in the box. Hint: For the first exercise color the first 5 blocks blue, and the next 5 blocks red.

$5 + \square = 10$

$3 + \square = 10$

> Great is our Lord, and mighty in power; His understanding is infinite. (Psalm 147:5)

So far you have only learned a little bit about God's world. But God knows everything. His understanding is so big. He knows so much that we cannot measure how much God knows!

God made everything. And God gave us math to help us to understand His world. He made numbers so we can count some of the stars. But we can't count all of them. There are too many for us to count!

God made addition and subtraction of big and small numbers. God made light, and He made the colors and the shapes around you. He made time and measurements and everything else you learned about in this book.

God helped you learn math. You prayed to Him when things were hard, and He heard you. God gave you the ability to learn and to practice your math.

Let's just thank God for these things today. Now, you can learn more of God's world. You can learn more of God's math too! Now, why not go on to discover more of what He has made! Build upon what you already know. Trust that He will help you as you learn and grow in knowledge. But most importantly, we learn more about God when we study His world.

Praise God for helping you through your first math book!

CERTIFICATION OF COMPLETION

CONGRATULATIONS!

You have completed *My First Numbers*.

You are now ready to go on to Math 1!

____

Name

____

Date Completed

____

Signature of Parent/Teacher

Answer Key

LESSON 2
Page 8 — 2, 1, 3; 2, 3, 1

LESSON 3
Page 9 — 0, 2, 1; 1, 3, 2
Page 10 — 2, 1, 3

LESSON 4
Page 12 — 5, 4

Page 14 — Answers will vary

LESSON 5
Page 16 — 4, 6, 5

LESSON 6
Page 17 — 4; 1, 2, 3, 4 ,5
Page 18 — 6; 5
Page 19 — G, O, D
Page 20 — 4, 1, 3, 6, 2; 7, 5

LESSON 7
Page 21 — 0, 6, 3, 2; 1, 4, 7

LESSON 8
Page 25 — Fourth; eighth; third; seventh; first; second; sixth; fifth

LESSON 9
Page 30 — Answers will vary
Page 31 — Answers will vary

LESSON 10
Page 32 — 4, 5, 2, 3, 8, 9
Page 33 — Images from left to right 4, 6, 2, 8; 7, 5; 10, 3, 6, 9.

LESSON 11
Page 40 — Red; Yellow; Blue
Page 40 — 5, 2, 3

LESSON 13
Page 44 — 2
Page 45 — 1; 2
Page 46 — 1; 3; 7

LESSON 14
Page 47 — 3; 5; 1
Page 48: 10; 8; 6

LESSON 15
Page 50 — 1, 1, 2; 2, 2, 4
Page 51 — 1, 2, 3; 2, 2, 4; 1, 0, 1; 1, 2, 3

LESSON 16
Page 53 — 1, 2, 3; 3, 1, 4;, 5; 1, 2, 3
Page 54 — 2; 4; 4; 3; 3
Page 55 — 4; 5

LESSON 17
Page 56 — 3; 4; 4; 5; 2
Page 57 — 5; 4; 2; 3; 3; 5
Page 58 — Answers will vary

LESSON 18
Page 59 — 2, 3, 5; 3, 2, 5; 5, 0, 5
Page 60 — Houses left to right 4, 5, 1, 3, 2

LESSON 19
Page 61 — Circle fire truck
Page 62 — Circle desktop computer; basketball; yo-yos; peaches

LESSON 20
Page 63 — Circle skateboard; lamp

LESSON 21
Page 65 — Watermelon; Pine tree; Pine tree; Heaviest: Pine tree & Lightest: Pea
Page 66 — 3, 4, 4, 5
Page 67 — Connect the dots from 1 to 10.

LESSON 22
Page 72 —

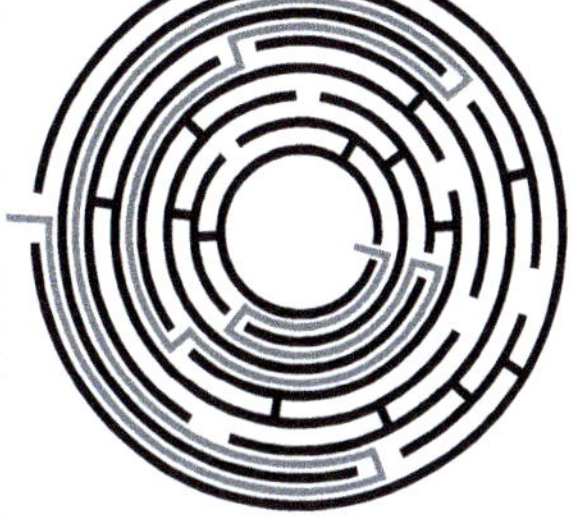

ANSWER KEY

LESSON 23
Page 73 — Answers may vary
Page 74 — 1, 4; 6, 3; 3, 5
Page 75 — Answers may vary

LESSON 24
Page 77 — 10; 10

LESSON 25
Page 79 — Answers will vary

LESSON 26
Page 80 — 2, 2, 4; 3, 2, 5; 1, 2, 3; 3, 1, 4
Page 81 — Color circles red, squares blue, triangles and rectangle green; 4, 7, 6, 5, 3
Page 82 — Pictures will vary. 2; 4; 3

LESSON 27
Page 84 — 1, 1, 2; 4, 11, 7; 9, 6, 7

LESSON 30
Page 90 — Square, circle, diamond, triangle, rectangle
Page 91 — Red 6, yellow 3, lt. green 5; brown 4, green 2, blue 7
Chapter 4

LESSON 31
Page 94 —

Page 95 —

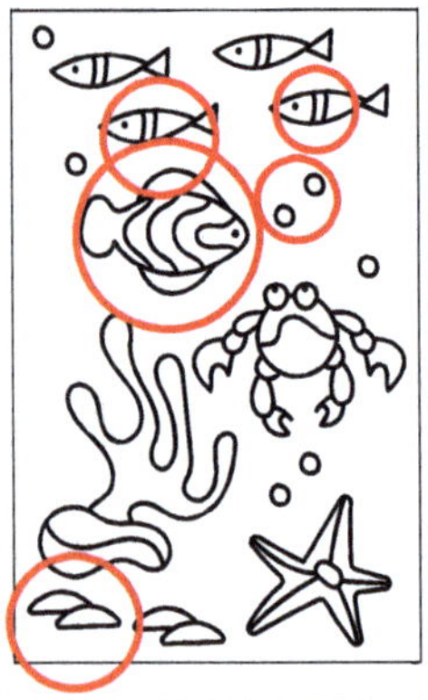

LESSON 32
Page 96 — 2, 2, 4; 1, 3, 4; 1, 2, 3
Page 97 — 1, 2, 3; 2, 3, 5; 3, 1, 4; 1, 3, 4

LESSON 33
Page 98 — 2, 3, 5, 1, 4

LESSON 34
Page 100 — 2, 3, 1, 4
Page 101 — 1, 4, 2, 3; Circle first; last

LESSON 35
Page 102 — 1, 2, 3, 4, 5, 6, 7; 2, 3, 1
Page 103 — 5, 1, 3, 6, 4, 2;
4; 2; 7; 3; 5; 1; 6

LESSON 36
Page 105 — 3, 3, 6; 5, 2, 7; 4, 5, 9; 1, 8, 9
Page 106 — 5; 6; 2, 4, 6, 9

LESSON 37
Page 107 — 4, 6, 9, 8, 10, 6, 9
Page 108 —

LESSON 38
Page 109 — 8; 6; 9
Page 110 — 10; 7; Connect dots from 1 to 10.

LESSON 39
Page 111 — 8, 8, 8, 9, 4, 5, 9, 10, 10, 7
Page 113 — 5, 1, 3

LESSON 42
Page 123 — 20, 10; 5, 10; 3, 5; 11, 10; (circle the 20, second 10, 5 and 11)
Page 124 — Missing coin is the last coin on the top row.

LESSON 43
Page 126 — 1, 2, 3; 1, 3, 4; 5, 2, 7; 5, 5, 10
Page 127 — 7, 8, 6, 6, 6, 6, 10, 7, 5, 9

LESSON 44
Page 129 — First, fourth, third; second, first, fifth; third, third, second; fourth, fifth, fourth; fifth, second, first
Page 130 — Banks from left to right 5, 10, 10, 5, 10.

LESSON 45
Page 132 — 1:00, 2:00, 3:00, 4:00, 5:00, 6:00, 7:00, 8:00, 9:00, 10:00, 11:00, 12:00

LESSON 46
Page 133 — 7, 3, 5, 4; 12, 10, 8, 1
Page 134 — Answers will vary

LESSON 47
Page 135 — Answers will vary

LESSON 48
Page 137 — 2; 6; 10; 15; 21 (Circle the 21.)
Page 138 — 4, 10, 12, 17, 14, 20, 6
Page 141 — 3, 10, 7, 2, 4, 11

LESSON 49
Page 144 — 11
Page 145 — 6; 2; 3; 8

LESSON 50
Page 146 — 4; 9; 7; 16; 25; 29

Page 147 — 2, 4, 5, 5; 3, 4, 2, 1

LESSON 51

Page 148 — 2, 6, 8, 5, 3, 9; Answers will vary
Page 149 — 8
Page 150 — Connect dots from 1 to 8.

LESSON 52

Page 152 — Top row 4, 2, 5, 3; Bottom row 3, 5, 2, 4

LESSON 53

Page 153 — 3, 1, 2, 5, 4; Answers will vary
Page 154 — 2, 7, 9; 2, 5, 7; 2, 8, 10; 2, 3, 5

LESSON 54

Page 156 — 8, 6, 1, 2, 3, 5
Page 157 — 7; 8; 9; 9; 10

LESSON 55

Page 158 — 2, 2; 1, 1 Circle 4 oranges, 5 oranges; 2 oranges, 5 oranges
Page 159 — Circle parrot, watermelon, books; cake, backpack, iron. Arrows on remaining items.

LESSON 56

Page 160 — 10, 3, 6, 7, 5
Page 161 — Answers will vary
Page 163 — 9, 10

LESSON 57

Page 164 — 3, 1, 3, 2; 3, 1, 3, 2, 9
Page 167 — Train 2, boat 3, plane 1, car 5
Page 168 — 5, 3, 4, 5, 6
Chapter 7

LESSON 58

Page 172 — 4
Page 173 — 7, 7, 5, 10
Page 174 — Connect dots from 1 to 15. 15, 8, 7

LESSON 59

Page 175 — 13, 12
Page 176 — 15, 12
Page 177 — 8, 8, 16; 9, 6, 15; 6, 7, 13; 5, 8, 13; 8, 4, 12

LESSON 60

Page 178 — 4, 3, 7; 1, 9, 10; 2, 6, 8
Page 179 — 3, 5, 8; 3, 4, 7; 5, 4 9; Answers will vary

LESSON 61

Page 181 — 10, 12; 11, 9; 13, 9; 4, 9; 15, 8; 10, 15
Page 182 — 12, 10, 11, 13, 12;
14, 11, 13, 10, 14

LESSON 62

Page 183 — 9, 9; 8, 4; 7, 7; 8, 8; 7, 9
Page 184 — 1, 1, 3; 2, 3, 4, 7; 3, 4, 5, 8; 4, 5, 6, 8

LESSON 63

Page 186 — 3, 1, 2; 4, 3, 1; 4, 2, 2
Page 187 — 2, 3, 2, 3

LESSON 64

Page 188 — 3, 1, 2, 1, 1
Page 189 — 2, 1, 1, 3

Page 190 —

LESSON 65

Page 191 — 2, 2, 1, 4
Page 192 — 2, 1, 3, 1, 1

LESSON 66

Page 193 — Blue; red; blue; blue
Page 194 — 3; 7; 2; 5; 8; 4; 1; 6

LESSON 67

Page 200 — Rain cloud; sun; sun; beetle; fly; butterfly

LESSON 68

Page 201 — Blue; red; blue; blue
Page 202 — Orange; blue; green; blue
Page 202 — Red; green; purple; green

LESSON 69

Page 203 — 9
Page 204 — 2, 4, 6, 7, 9;
3, 6, 8, 10; 8, 7; 5, 4, 5; 1, 2, 1
Page 205 — 2, 4; 9, 11; 12, 14; 5, 7; 8, 10; 15, 17; 18, 20; 7, 9; 10, 12

LESSON 70

Page 206 — 8, 10, 12; 12, 15, 18; 16, 17, 18
Page 207 — Tree, map, pine tree; monkey, cat, giraffe

LESSON 71

Page 210 — 5; 2; 5; 6
Page 211 —

LESSON 72

Page 212 — 6, 2, 4; 10, 2, 8
Page 213 — 5, 1, 4; 3, 7, 2, 6; 9, 8

LESSON 73

Page 214 — 4, 2; 2, 1; 3, 7; 4, 3; 0, 3
Page 215 — 7, 4, 3; 6, 2, 4; 9, 4, 5
Page 216 — 2; 4; 2; 4, 6, 5; 7, 5, 6; 3, 2, 2; 5, 3, 2

ANSWER KEY

LESSON 74

Page 217 — 5, 5, 2

Page 218 —

Page 219 — Answers on balls from left to right
9, 1, 7, 10, 8, 9;
4, 5, 6, 4;
3, 2, 1

LESSON 75

Page 223 — 5, 50

Page 224 — 20, 30, 60, 40

LESSON 76

Page 226 — 60, 20, 90, 40, 60, 80

Page 227 — 20, 40, 60, 80, 90; Connect the dots from 10 to 90 counting by 10.

LESSON 77

Page 229 — 2, 3, 4, 5, 3, 4, 5, 6

Page 230 — 1, 2, 3, 4, 0, 1, 2, 3

LESSON 78

Page 231 — 4, 20

Page 233 — 10

LESSON 79

Page 234 — 16, 13, 14, 20, 30, 17

Page 235 — 7, 6, 9, 10

LESSON 80

Page 236 — 10, 20, 25, 35, 40, 50

Page 237 — 30, 50, 70

LESSON 81

Page 238 — 7, 3, 10; 4, 2 6

Page 239 — 5, 5, 10; 2, 5, 7; 8, 2; 10

Page 240 — 9, 12; 10, 13; 11, 21; 8, 20; 15, 30; 10, 40

LESSON 82

Page 241 — 9; 8; 9; 10

Page 242 —

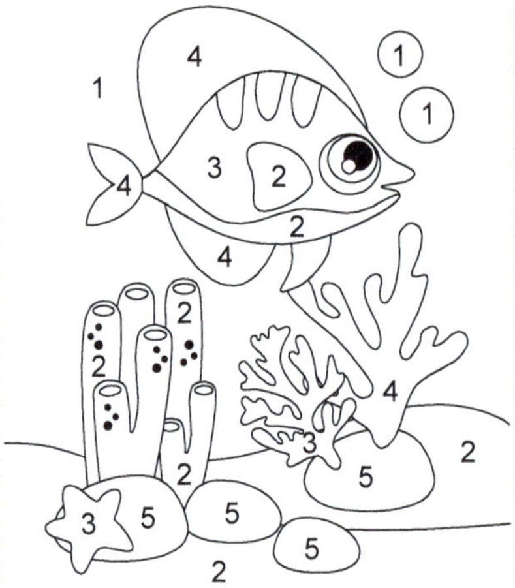

LESSON 83

Page 243 — 17; 20; 16; 13; 15

Page 244 — Circle 3; 7; 7; 9

LESSON 84

Page 249 — 18, 14; 17, 8; 15, 7; 16; 14, 7; 20, 13; 12, 0

LESSON 85

Page 250 — 9, 9

Page 251 — 8, 5, 7, 10

LESSON 86

Page 253 — 40, 30; 80, 60

Page 254 — 5; 17; 23; 32; 48; 54; 70; 79; 83; 95

LESSON 87

Page 256 — 2, 3, 5; 4, 1, 3

Page 257 — 6, 2, 4; 4, 2, 2; 0, 17, 74, 38, 6, 13

Page 258 — Connect dots from 1 to 50.

LESSON 88

Page 259 — Third, fourth, first, second.

Page 260 — 6, 7, 4, 5, 3, 2

Page 261 — 4, 6, 10; 1, 5, 6; 4, 4, 8; 7, 3, 10

LESSON 89

Page 263 — Circle banana, orange, sandwich.

LESSON 90

Page 264 — Circle 4; triangle 4; square 4.

Page 265 —

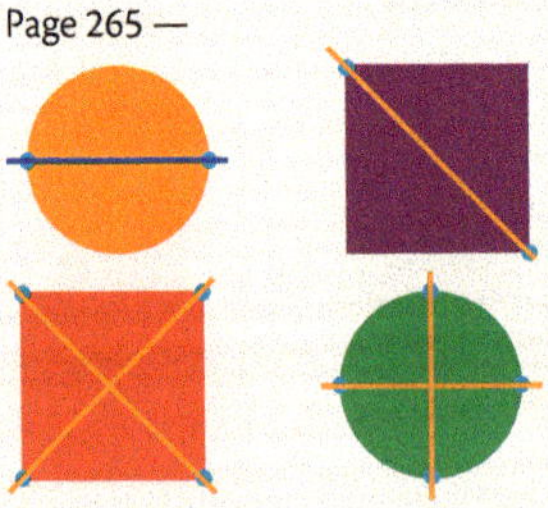

Page 266 — Circle the last in second row and middle of third row.

LESSON 91

Page 267 — 4, 8, 3, 2

Page 268 — 8, 5, 3; 7, 4, 3

Page 269 — 4, 3, 1; 5, 7

Manipulative Cutouts

Make ten copies of this page and cut out the manipulatives.

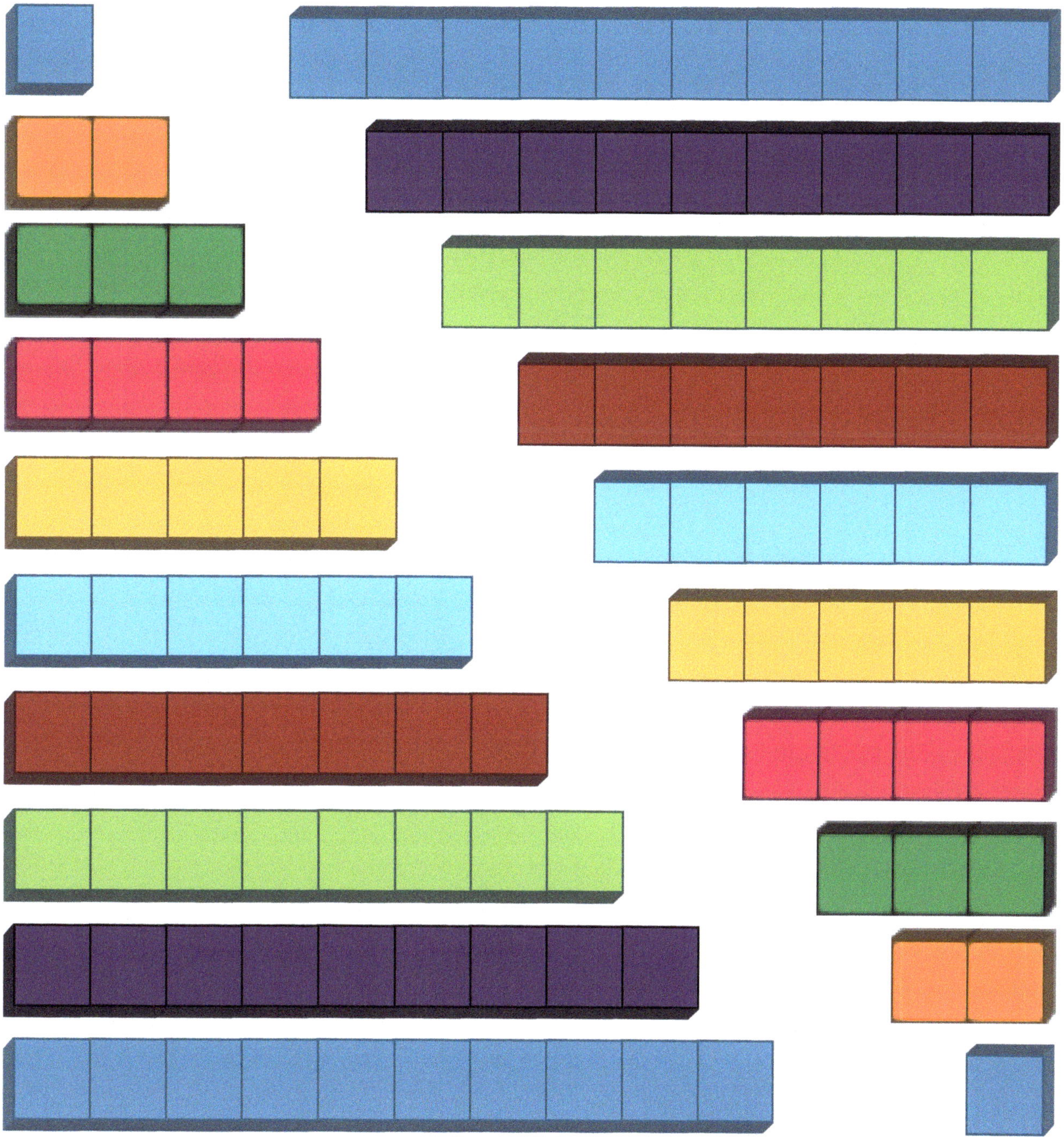

THIS PAGE IS LEFT INTENTIONALLY BLANK

1-10 Number Line

1-20 Number Line

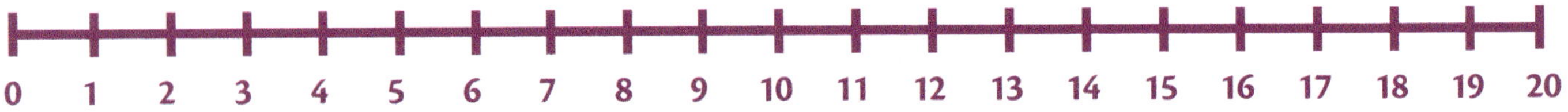

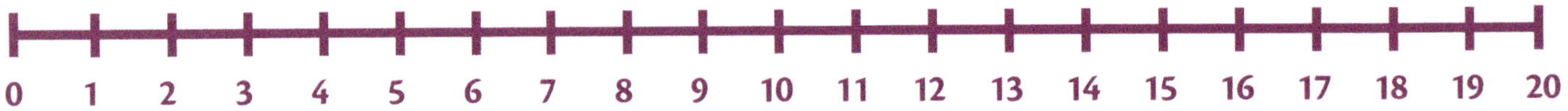

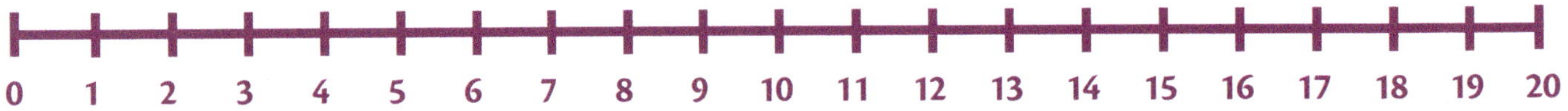

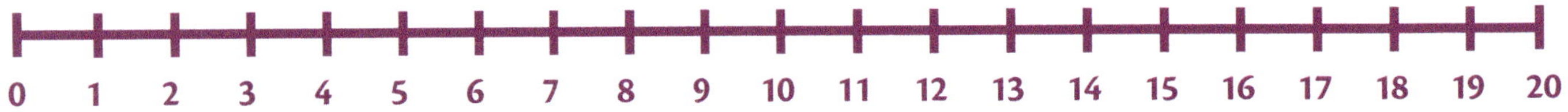

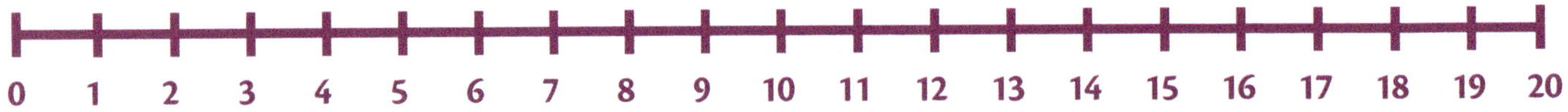

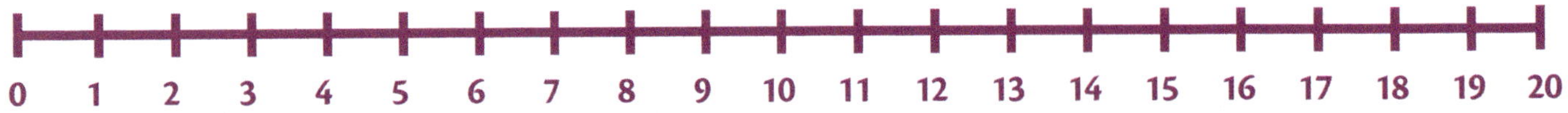

Shapes and Colors

Make It! (page 90)

THIS PAGE IS LEFT INTENTIONALLY BLANK

Shape Games

Challenge Game #1 (page 88)

Challenge Game #2

THIS PAGE IS LEFT INTENTIONALLY BLANK

0+1

9

0+2

8

0+3

7

1+1

6

1+2

5

1+3

4

1+4

7

1+5

8

1+6	**1+7**	**1+8**
3	2	1
2+2	**2+3**	**2+4**
4	3	2

2+6	**2+5**
6	5

2+7

8

3+1

7

3+2

6

3+3

7

3+4

6

3+5

9

3+6

9

4+1

8

4+2

5

4+3

4

4+4

9

4+5

8

5+1

7

5+2

6

5+3

5

5+4

9

6+1

4

6+2

3

6+3

2

7+1

7

7+2

6

8+1

5

1-1

0

2-1

8

3-1

9

4-1

8

5-1

7

6-1

9

7-1

9

8-1

8

9-1

1

2-2

0

3-2

3

4-2

2

5-2

1

6-2

6

7-2

5

8-2

4

9-2

1

3-3

0

4-3

3

5-3

2

6-3

1

7-3

6

8-3

5

9-3

4

4-4

0

5-4

1

6-4

0

7-4

2

8-4

1

9-4

1

5-5

2

6-5

0

7-5

1

7-6

4

8-6

3

7-7

2

6-6

1

9-7

0

8-7

5

9-8

1